Synthesis Lectures on Mathematics & Statistics

Series Editor

Steven G. Krantz, Department of Mathematics, Washington University, Saint Louis, USA

This series includes titles in applied mathematics and statistics for cross-disciplinary STEM professionals, educators, researchers, and students. The series focuses on new and traditional techniques to develop mathematical knowledge and skills, an understanding of core mathematical reasoning, and the ability to utilize data in specific applications.

Lih-Yuan Deng · Nirman Kumar ·
Henry Horng-Shing Lu · Ching-Chi Yang

Random Number Generators for Computer Simulation and Cyber Security

Design, Search, Theory, and Application

Springer

Lih-Yuan Deng
Department of Mathematical Sciences
University of Memphis
Memphis, TN, USA

Henry Horng-Shing Lu
Biomedical Artificial Intelligence Academy
Kaohsiung Medical University
Kaohsiung, Taiwan

Department of Medical Research
Kaohsiung Medical University Hospital
Kaohsiung, Taiwan

Institute of Statistics
National Yang Ming Chiao Tung University
Hsinchu, Taiwan

Department of Statistics and Data Science
Cornell University
Ithaca, New York, USA

Nirman Kumar
Oracle America, Inc
Austin, TX, USA

Ching-Chi Yang
Department of Mathematical Sciences
University of Memphis
Memphis, TN, USA

ISSN 1938-1743   ISSN 1938-1751   (electronic)
Synthesis Lectures on Mathematics & Statistics
ISBN 978-3-031-76721-0   ISBN 978-3-031-76722-7   (eBook)
https://doi.org/10.1007/978-3-031-76722-7

This Springer imprint is published by the registered company Springer Nature Switzerland AG
The registered company address is: Gewerbestrasse 11, 6330 Cham, Switzerland

If disposing of this product, please recycle the paper.

Preface

Random number generators have two major areas of application: (1) computer simulation and (2) computer security. The random number generators are utilized in scientific simulation applications such as random sampling from a population, bootstrap resampling methods, and MCMC (Markov Chain Monte Carlo) methods. More broadly, simulations are commonly used to evaluate and compare complex stochastic processes or procedures where analytical solutions are infeasible.

In computer security, random number generation is crucial for applications such as digital signature schemes, password generation, online gambling/transactions, cryptography, and online security. Until recently, a significant divide existed between these two areas of application due to differing requirements.

Effective random number generators for computer simulation are efficient, possess excellent high-dimensional properties, have long period lengths, and perform well empirically. Recent advancements have led to the design and search for generators with extremely long period lengths and strong statistical justifications. However, many of these proposed generators are linear and can be easily predicted from a few past observations.

Conversely, the primary requirements for random number generators in computer security are unpredictability, uniformity, efficiency, and randomness. Classical generators often relied on theoretical properties or conjectures, such as the difficulty of factoring very large integers (e.g., 200 digits). As expected, such algorithms are highly inefficient with unknown empirical performance. Modern *secure* generators use a complex series of ARX (Addition, Rotation, and Exclusive OR) operations to form *internal states*, initialized from a given key and initialization values using a key expansion scheme. These generators often lack strong theoretical support for properties such as uniformity, period length, and randomness. Achieving these desirable properties of uniformity and randomness typically requires many rounds of ARX operations, potentially slowing down their generating speed.

In this book, we promote the general procedure of combining classical generators with proven distributional properties and secure generators with tested cryptanalysis. The goal is to inherit the strengths of both: the great distributional properties of classical linear generators and the proven security property of secure generators. This new class of generators should be suitable for both computer simulations and cybersecurity applications.

This book is intended to serve as both a reference and a textbook. It can be used as the main textbook or supplementary material for various courses at the graduate or advanced undergraduate level. A deep background in mathematics, statistics, cryptography, or computer science is not required. Basic knowledge of number theory, statistical theory, and scientific computer literacy will be helpful but not necessary. We do not rely on a particular software system in the book, though C/C++ and the availability of S-Plus or R could be useful.

General courses in statistical computing or computational statistics could start Part I of the textbook, which discusses classical generators for computer simulation applications and recent developments. Several issues of design and computer search algorithms with statistical justifications are also covered. Part II focuses on the quality assessment of random number generators, including new efficient algorithms to improve the classical spectral test. It also includes a discussion and empirical evaluation of the classical built-in generators in the popular R package. Part III addresses the parallelization of these generators, which may require a deeper mathematical background. General courses in cryptography could use this book as supplementary material, covering the first few chapters of Part I and then concentrating on Part IV, which deals with combining both methods to create new ciphers suitable for both computer simulation and cybersecurity applications.

Memphis, TN, USA   Lih-Yuan Deng
Austin, TX, USA   Nirman Kumar
Kaohsiung, Taiwan   Henry Horng-Shing Lu
Memphis, TN, USA   Ching-Chi Yang
June 2024

Acknowledgements

This book would not be possible without the close collaborations with all co-authors during many weeks of Zoom meetings in the past year. In addition, Lih-Yuan Deng would like to thank his Ph.D. students working on the topic of random number generation under his supervision (in chronological order): Yu-Chao Chu, Yilian Yuan, Gwei-Hung Tsai, Jiannnong Wang, Huajiang Li, Bryan Winter, Kenneth Bobvah Pasiah, Farnaz Solatikia, and Jonathan Robert McCurdy. They have provided key sources of research inspiration and motivation with constant challenges and insightful discussions. In addition, he would also like to thank all of his co-authors for their help in getting these papers published.

Lih-Yuan Deng acknowledges support from the National Science Foundation under an award for The Learner Data Institute (NSF-1934745: The opinions, findings, and results are solely the authors' and do not reflect those of the funding agency.) This book has been completed during his sabbatical leave with generous support from the University of Memphis as well as the hosting National Yang Ming Chiao Tung University.

Nirman Kumar would like to thank Lih-Yuan Deng and Ching-Chi Yang for the opportunity to collaborate on this book and other research projects, and the learning experience it provided. He would also like to thank his family for the constant support they provided throughout the project.

Horng-Shing Lu gratefully acknowledges the support of the National Science and Technology Council, Taiwan, ROC, through Grant No. 112-2634-F-A49-003 and 110-2118-M-A49-002-MY3.

Ching-Chi Yang would like to express the deepest gratitude to his co-authors and collaborators, whose invaluable insights and tireless efforts have made this book possible. He is particularly indebted to Lih-Yuan Deng (his mentor) and Dennis K. J. Lin (his advisor) who have guided him with their wisdom and expertise.

Finally, we would also like to thank Katherine McKinney Link for her invaluable assistance in proofreading and making this book possible.

Contents

Part I
Classical Random Number Generators

Introduction

In most fields of scientific research today, it is common to use computer simulation to evaluate statistical methodologies, to take random samples from target distributions, and to use Markov Chain Monte Carlo (MCMC) for simulating posterior distributions. The National Air Force Research Laboratories report that a large portion of supercomputer resources are currently spent on Monte Carlo computations (see, e.g., https://www.kirtland.af.mil/News/Article-Display/Article/3497652/). The results of a computer-intensive study based on random sampling depend largely on the quality of (pseudo)-random numbers available, which in turn depends on the properties of the pseudo-random number generator used. Typically, to produce a sequence of random variables from a specified distribution, a *deterministic process* generates a sequence of random integers between 0 and some integer m which are then transformed into variates in [0, 1] by dividing by m. These "uniform" variates are further transformed into variables with the desired distribution. This deterministic process typically starts with some "randomly chosen" seeds and is known as a pseudo-random number generator (PRNG). Several commonly used PRNGs are presented in this book.

There are typically five major issues considered when selecting or comparing pseudo-random number generators. These are, (1) high period length, (2) uniformity and the equidistribution property, (3) computing efficiency and portability, (4) statistical justifications and empirical performance, and (5) ability to parallelize and generate substreams. All such PRNGs are periodic sequences and eventually repeat after some period length which should be large for the PRNG to be useful. The equidistribution property up to t dimensions ensures that virtually every d-tuple ($d \leq t$) of integers appears exactly the same number of times over the entire period of the generator. The properties of equidistribution over high-dimension spaces and longer period length can be achieved by selecting a class of generators that have these properties. Computing efficiency and portability are then achieved by carefully chosen

L. Deng et al., *Random Number Generators for Computer Simulation and Cyber Security*, Synthesis Lectures on Mathematics & Statistics, https://doi.org/10.1007/978-3-031-76722-7_1

designs of these generators. The empirical properties and statistical justifications can inspire the development of ever better PRNGs with stronger empirical performances. The issues of parallelization are becoming increasingly more important with the increasing use and access to ever more powerful computing resources.

The use of pseudo-random number generators on single processors has been well studied, but less work has been done in the case of parallel computing. Parallel computing is a common way to speed up computations by breaking a task into smaller relatively independent subtasks that can be handled by individual processors in parallel. In this case, in order to obtain good overall results, it is necessary that good parallel random number generators be used. The design and implementation of suitable and independent parallel PRNGs are also discussed in this book.

A tremendous amount of computer software has been developed to support statistical computing requirements in the past decade. Simulation study has become a common practice in the development of statistical methods. This is particularly true when the analytical study is intractable. The generation of random variates with a specified distribution is a key issue of almost any simulation study. Given the distribution, often there are several generating methods that can be used to produce a random number sequence. These methods are mainly based on the generation of independent variates from the uniform distribution, $U(0, 1)$. Thus, uniform random number generation is the foundation for all computer simulation studies. Because of this, we will focus on $U(0, 1)$ random number generators in this book. Note that random numbers generated by any specific algorithm are "systematic" and therefore are neither truly independent, nor purely random.

Many random number generators have been proposed in the literature. The most popular generator, the *Linear Congruential Generator* (LCG), was proposed by Lehmer (1951) and is widely used today. It is generated from a simple linear recursive relation with a modulus p, which is commonly chosen as a large prime number or of the form 2^w to fit in the computer word size (say, 32-bits). Once pseudo-random integers between 0 and p are generated, they are transformed into the interval $[0, 1]$ by scaling by $1/p$. One of the major limitations of the LCG is that its period is limited by the modulus p. Most software uses $p = 2^{31} - 1$. This may be sufficiently large for most simulations done in the past, however, because better and better computer facilities are now available, the scale of the simulation study is getting larger and larger. For a large-scale simulation study, such a life period may not be enough. This is particularly true for the parallel simulation systems.

1.1 Recent Developments

Chapter 2 briefly discusses several classical PRNGs for computer simulations and their basic properties. In particular, Multiple Recursive Generator (MRG) and Matrix Congruential Generator (MCG) are two important extensions to the Classical Congruential Generator (LCG). Both MRG and MCG of a larger order have a large period length and equidistribution property over a high dimension. However, there are several problems, and there are developments on possible solutions:

1. The classical search algorithm, Algorithm AK, to find such generators with maximal period length, is highly inefficient when the order k is large. We will discuss its solution, Algorithm GMP, in Chap. 4. The main improvements are the early exit strategy and the use of primality checking in place of the more challenging problem of factorization.

2. The general form of large-order MRGs is computationally inefficient because a large number of multiplications are needed to compute its next variate. We will discuss several special forms of MRGs which can be as efficient as LCGs. In particular, DX generators will set lots of multipliers to be zero and set coefficients for non-zero multipliers to be the same to minimize the multiplication cost. There are other efficient generators proposed in Chap. 3.

3. The issue of parallelization of PRNGs is an important issue for both LCGs and large-order MRGs. A traditional parallelization procedure is to divide the whole period into a number of disjoint blocks with a fixed block size. There are two potential problems with this approach: (a) no efficient algorithm for jumping ahead, especially for large-order MRGs, and (b) the same equations for each block may yield a simple linear structure/relationship, especially for LCG. Chapter 8 discusses a systematic random leapfrog method for LCG which will yield every block with different LCGs each with maximum period length. Extending this method for LCG, Chap. 8 also discusses AGM methods for large-order DX generators to automatically produce many MRGs with few non-zero terms. Chapters 9 and 10 consider the problem of parallelization via the relationship between MRG and MCG.

4. Another issue is the comparison among LCGs and MRGs of the same order/structure. It is well-known to use the "gold standard" spectral test to evaluate the "goodness" of the LCG which is the "maximum gap" between the parallel line/hyperplanes covering the generated sequence. The traditional algorithm is not very efficient even for LCG spectral test. To illustrate an efficient algorithm, in Chap. 6, we use LCG as an example to compare the traditional method and a newly proposed LCG spectral test. In Chap. 7, this efficient procedure can then be extended for large-order MRGs, especially for DX generators. Since it is much easier to check the spectral test property than its maximum period length, we can use the spectral test first and then check for the maximum period length.

5. Classical PRNGs are suitable for computer simulations because of the period length, high-dimensional equation, parallelization, and performance on empirical tests. However, they are unsuitable for computer security applications mainly because of their simple and linear structure. Therefore, most of the proposed secure ciphers avoid using classical linear generators. In Chap. 12, we describe all four finalists in Profile 1 (software ciphers) and three finalists in Profile 2 (hardware ciphers) in the popular eSTREAM project. To ensure the "randomness" and "uniformity" of the generated sequences, most

of them use a long series of ARX (Arithmetic, Rotation, and eXclusive or) operations for the initialization and key generation of some internal state vectors. In Chap. 13, we discuss several common methods to improve the security property: ARX, shuffling, mutual shuffling, and linear generators as baseline generators. Chapter 14 offers an example of mutual shuffling for the SAFE (Secure And Fast Encryption) procedure with two good external number generators. Chapter 15 describes adding one external generator to the classical RC4. Likewise, Chaps. 16, 17 and 18 discuss the addition of external generators as extensions/enhancements to several of the eSTREAM Profile 1 software cipher finalists.

1.2   Outline of the Book

This book is organized in 4 parts. Part I (Chaps. 1–3) discusses the most popular linear generators including Linear Congruential Generators (LCGs) and their generalizations, Multiple Recursive Generators (MRGs), and Matrix Congruential Generators (MCGs). We also discuss Linear Feedback Shift Registers (LFSRs) and Generalized Feedback Shift Registers (GFSRs) which are more efficient but restricted (in choice of some parameters) versions of MRGs. The currently most popular generator is MT19937 which can be considered a special case GFSR. Large-order MRGs have several nice properties with long period length, high dimensional equidistribution property, and great empirical performances. Part I also addresses potential problems of larger order MRGs: (a) design of efficient large MRGs in Chap. 3 such as the DX-k-s generators and (b) the need for an efficient search algorithm in Chap. 4. Chapter 5 discusses generic ways to improve basic generators (as per one or more of the judgment criteria listed above), and mathematical and statistical justification for why such methods work. We also discuss some built-in generators in R, arguably the most popular programming language for statistical applications.

Part II (Chaps. 6 and 7) of the book discusses the quality assessments of the linear generators including LCG and large-order MRGs. There are two assessments: (a) theoretical spectral test (b) empirical test like TestU01. Spectral test is a "gold standard" to evaluate the goodness of LCGs, but it is unclear if this extends to the evaluation of large-order MRGs. To motivate the new algorithm, Chap. 6 compares the difference between the new method and the classical method for the LCG spectral test. Chapter 7 extends this idea to compute spectral tests for large-order MRGs, especially for DX generators.

Part III (Chaps. 8–10) discusses the issue of parallelization of the large-order MRGs and LCGs. The classical method is to break the whole period length into several blocks of equal block size with some efficient jump scheme. Such a scheme is straightforward for LCGs or for very small-order MRGs. In Chap. 8, we discuss the issue for LCG with different ways to break the whole period length of LCG using the systematic random leapfrog method. Further extension of this idea can be used for DX generators. Using the relationship between MRGs and MCGs, Chaps. 9 and 10 discuss the efficient implementation/parallelization of MRG vis MCG.

Part IV (Chaps. 11–18) discusses the need and requirement of secure random number generators. Classical linear generators are great for computer simulation applications but not for computer security applications mainly because of their linearity structure. Therefore, most of the secure generators/ciphers such as the one proposed in eSTREAM project (discussed in Chap. 12) do not use any classical linear generators. Without using these linear generators, they have to rely on a long series of non-linear operations to achieve the "randomness" and "uniformity" of generated sequences. No such "theoretical justification" can be provided except, hopefully, such non-linear operations are hard to "reconstruct" by observing a sequence of variates. Chapter 13 discusses various methods to improve the security of the generators, including mutual shuffling and adding external generators. Several examples of possible enhancements to the existing secure ciphers particularly with additions of good external generators and shown in various chapters in Part IV.

1.3 Some Basic Concepts and Definitions

In this section, we define some preliminary terms that find their use throughout the book.

Definition 1.1 ($\mathbb{Z}_m$) The set $\{0, 1, \ldots, m-1\}$ along with the addition and multiplication operations modulo an integer $m \geq 1$ is the ring $\mathbb{Z}_m$. If $m = p$, a prime, this ring is also a field.

Definition 1.2 (*PRNG*) A random number generator (PRNG) is a sequence $X_{-k}, X_{-(k-1)}, \ldots, X_{-1}, X_0, X_1, \ldots$, where k is a fixed integer and for some function $f : \mathbb{Z}_p^k \to \mathbb{Z}_p$,

$$X_i = f(X_{i-1}, \ldots, X_{i-k}) \mod p, \quad i \geq 0, \tag{1.1}$$

and $X_{-k}, X_{-(k-1)}, \ldots, X_{-1}$ are initial seeds taken from $\mathbb{Z}_p = \{0, 1, \ldots, p-1\}$. Generally p is a prime number.

Given a PRNG as above the variates X_i/p are in $[0, 1]$ and (are expected to) behave as samples of a random variable $U \sim U(0, 1)$. Clearly, this is not true since X_i/p can only assume p distinct values. However, if p is very large, the set of assumed values is sufficiently dense in $[0, 1]$ and for many practical purposes (such as simulation applications) will be good enough.

Definition 1.3 (*Period of a PRNG*) The period of a PRNG is the minimum integer $r > 0$ such that $X_i = X_{i+r}$ for all $i \geq -k$.

Classical Random Number Generators for Computer Simulation

Computer simulation plays a prominent role in contemporary scientific research across various disciplines. It is widely employed to evaluate statistical methodologies, derive random samples from target distributions, and utilize the Markov Chain Monte Carlo (MCMC) technique for simulating posterior distributions.

When it comes to generating random numbers, there exist multiple methods, including Inverse Transform Sampling, Rejection Sampling, Markov Chain Monte Carlo, and the Metropolis-Hastings algorithm. Our focus will primarily be on Inverse Transform Sampling as in Example 2.1, with the acknowledgment that subsequent methods rely on established distributions to generate unknown or intricate distributions.

Example 2.1 (*Inverse Transform (CDF) Sampling*)

Inverse Transform Sampling, a technique employed in generating random numbers, hinges on the concept of Cumulative Distribution Function (CDF). The procedure involves the transformation of uniform variates into variables adhering to the desired distribution.

Suppose Y is the random variable of interest and its cumulative distribution function is known, denoted by $F(y)$. If one generates a series of i.i.d. samples $\{x_i, i = 1, 2, 3, ...\}$ from a uniform distribution between 0 and 1, the values $F^{-1}(x_i)$ will follow the distribution of Y.

A *pseudo-random number generator* (RNG) is a deterministic algorithm that uses a starting seed to generate a sequence of numbers that *look* like a sequence of uniform, random, and independent numbers from some set (see Definition 1.2). In other words, they enjoy properties similar to a sequence of numbers chosen uniformly at random from the set. Such a RNG sequence is only resistant to detection by a class of algorithms but not arbitrary algorithms (for example the one generating it), and the more resistant it is, the better it is. For our purposes, we will say it is good if common statistical tests cannot distinguish the RNG sequence from a truly random sequence. This chapter introduces several frequently used RNGs. In essence, the crux of generating pseudo-random numbers lies in the uniform random number generator U(0,1).

Uniform Random Number Generators

The success of computer-intensive studies reliant on random sampling hinges on the quality of available (pseudo) random numbers. This, in turn, is contingent upon the characteristics of the pseudo-random number generator in use. Typically, the process of generating a sequence of random variables that adhere to a given distribution involves generating independent random integers within the range of 0 to a specified integer value denoted as "p". These integers are subsequently transformed into variates within the interval [0,1] through division by "p".

We consider a general form of a pseudo-random number generator (RNG) as $X_{-k}, X_{-(k-1)}, \cdots, X_{-1}, X_0, X_1, \ldots$, where

$$X_i = f(X_{i-1}, \ldots, X_{i-k}) \bmod p, \quad i \geq 0, \tag{2.1}$$

$f : \mathbb{Z}_p^k \to \mathbb{Z}_p$ is a function of the most recent k integers in the past, and $X_{-k}, X_{-(k-1)}, \ldots, X_{-1}$ are initial seeds taken from $\mathbb{Z}_p = \{0, 1, \ldots, p-1\}$. Most RNGs are of this general form with some appropriate choice of function f.

For many computer applications, the goal of a RNG is to produce a sequence of variates that is very hard to distinguish from a sequence of (truly) random numbers. Therefore, an ideal RNG should satisfy the property of **High-dimensional equidistribution** (see Definition 21), **Efficiency**, **Long period length** and **Portability**, also known as the **HELP** property in Deng (2005).

2.1   Linear Congruential Generators (LCG)

A linear congruential generator (LCG), first proposed by Lehmer (1951), is the generator (Eq. 2.1) where f is a linear function with $k = 1$. Its sequence is obtained by

$$X_i = (BX_{i-1} + A) \bmod m, \quad i \geq 0, \tag{2.2}$$

where X_i, A, B, and p are non-negative integers and $X_{-1} \neq 0$ is chosen from $\mathbb{Z}_p$ as a seed. It is a simple and efficient algorithm designed to produce sequences of pseudo-random numbers that exhibit characteristics similar to those of truly random numbers. LCGs are particularly valuable in applications where statistical randomness is sufficient and computational speed is critical. The sequence is periodic, and if $A \neq 0$, it is possible to achieve the full period p.

The fundamental idea behind LCGs is to generate a sequence of numbers by repeatedly applying a linear congruential formula to the previous number in the sequence. However, according to Marsaglia (1972), the period (see Definition 1.3) cannot be greater than the period of the corresponding LCG with $A = 0$ as

$$X_i = B X_{i-1} \bmod m, \quad i \geq 0, \tag{2.3}$$

where $X_{-1} \neq 0$. The maximum period of the sequence $\{X_0, X_1, X_2, \ldots\}$ generated by LCGs in Eq. 2.3 depends on the choice of the modulus m (Knuth 1998, Theorem C, p. 20):

1. $m = 2^t, t > 3$. The maximum period attainable is 2^{t-2}.
2. $m = p > 3$, a large prime number. The maximum period attainable is $p - 1$.
3. When p is a prime number and B is a primitive element (also known as primitive root) modulo p, the LCG has a period of $p - 1$.

Note that B is a primitive root modulo p, if for any prime factor q of $(p - 1)$, $B^{(p-1)/q} \neq 1 \bmod p$. In other words, the smallest positive integer r with $B^r \bmod p = 1$ is $r = p - 1$. Notice that by Fermat's theorem, $B^{p-1} \bmod p = 1$ if B is not divisible by p. The following theorem from number theory characterizes all the primitive roots modulo p.

Theorem 21 *Let p be a prime number. If B is a primitive root modulo p, then $B^r \bmod p$ is also a primitive root if* $\gcd(r, p - 1) = 1$, *i.e., r is relatively prime to $p - 1$. Moreover, $B^r \bmod p$ for r with* $\gcd(r, p - 1) = 1$ *generate all the primitive roots modulo p.*

Hence, the total number of primitive roots is $\phi(p - 1)$, where $\phi(x)$ is the Euler's totient function counting the number of integers between 1 and x that are relatively prime to x. Deng et al. (1994) used this property to develop a procedure to automatically produce many maximum-period LCGs for parallel simulation to be discussed later.

Example 2.2 (*Primitive roots*)

Table 2.1 shows a simple example of primitive roots modulo $p = 7$. A primitive root is for example $B = 3$ since $B^r \bmod p$ is 1 for $r = p - 1$ but not so for any smaller r. Moreover, of the powers of B, the $B^1 \bmod p$ and $B^5 \bmod p$, i.e., for all the r with $\gcd(r, 6) = 1$, are primitive roots. (Larger values of r do not matter, as $B^1, \ldots, B^6 \bmod p$ generate all the possible remainders that repeat cyclically, i.e., $B^r \bmod p = B^{r+p-1} \bmod p$.)

Table 2.1 LCG (B, p = 7)

B	B^2	B^3	B^4	B^5	B^6
2	4	1	2	4	1
3	2	6	4	5	1
4	2	1	4	2	1
5	4	6	2	3	1
6	1	6	1	6	1

Example 2.3 (*Sophie Germain prime*)

Table 2.2 shows an example of Sophie Germain primitive roots modulo $p = 23$.

- For $p = 23$, maximum period with $B = 5, 7, 10, 11, 14, 15, 17, 19, 20, 21$.
- LCG period can only be a factor of $22(p - 1)$, i.e., 2, 11, 22.

For the case of $p = 2^{31} - 1$, 7 is the smallest primitive element; but $B = 7$ is not a suitable multiplier for LCG because it is too small (Knuth 1998). A minimum standard was proposed by Park and Miller (1988) by choosing an LCG with $p = 2^{31} - 1$ and $B = 7^5$, where 5 is chosen because it is the smallest exponent $r > 1$ with $\gcd(r, p - 1) = 1$. This LCG has been used in most computer systems and packages and has period $p - 1 \approx 2.1 \times 10^9$. To further speed up the generating efficiency, one can consider a special form of the multiplier $B = 2^r \pm 2^w$ for LCGs with $p = 2^{31} - 1$. A multiplier of this form, called powers-of-two decomposition, was first suggested by Wu (1997) for LCGs. It can result in faster computation because one can replace more expensive multiplication and modulus operations with more efficient logical shift and addition operations. Multipliers with powers-of-two decomposition can be useful to increase the generating efficiency for other RNGs to be discussed later.

There is a simple linear relationship between two successive variates produced by an LCG in Eq. 2.3. This has some serious implications about the "randomness" of the sequence. For example consider all the possible successive d-tuples $(X_i, X_{i+1}, \ldots, X_{i+d-1})$ where d is a fixed integer. All of these are points in the d-dimensional hypercube $\mathbb{Z}_p^d$. Marsaglia (1968) was the first to show that these successive overlapping sequences of d numbers fall on at most $(d!p)^{1/d}$ planes. This is roughly proportional to $p^{1/d}$ so the randomness gets "worse" with larger d. For example, Marsaglia (1968) points out that this can be much smaller than the minimum number of hyperplanes needed to cover all the lattice points in $\mathbb{Z}_p^d$ as would be expected of a truly random sequence generated as the set of all possible successive d-tuples. Even when d is of moderate size, successive d-tuples in the output sequence generated by the LCG will lie in a simple lattice structure.

Table 2.2 LCG (B, p = 23)

B	B^2	B^3	B^4	B^5	B^6	B^7	B^8	B^9	B^{10}	B^{11}	B^{12}	B^{13}	B^{14}	B^{15}	B^{16}	B^{17}	B^{18}	B^{19}	B^{20}	B^{21}	B^{22}
2	4	8	16	9	18	13	3	6	12	1	2	4	8	16	9	18	13	3	6	12	1
3	9	4	12	13	16	2	6	18	8	1	3	9	4	12	13	16	2	6	18	8	1
4	16	18	3	12	2	8	9	13	6	1	4	16	18	3	12	2	8	9	13	6	1
5	2	10	4	20	8	17	16	11	9	22	18	21	13	19	3	15	6	7	12	14	1
6	13	9	8	2	12	3	18	16	4	1	6	13	9	8	2	12	3	18	16	4	1
7	3	21	9	17	4	5	12	15	13	22	16	20	2	14	6	19	18	11	8	10	1
8	18	6	2	16	13	12	4	9	3	1	8	18	6	2	16	13	12	4	9	3	1
9	12	16	6	8	3	4	13	2	18	1	9	12	16	6	8	3	4	13	2	18	1
10	8	11	18	19	6	14	2	20	16	22	13	15	12	5	4	17	9	21	3	7	1
11	6	20	13	5	9	7	8	19	2	22	12	17	3	10	18	14	16	15	4	21	1
12	6	3	13	18	9	16	8	4	2	1	12	6	3	13	18	9	16	8	4	2	1
13	8	12	18	4	6	9	2	3	16	1	13	8	12	18	4	6	9	2	3	16	1
14	12	7	6	15	3	19	13	21	18	22	9	11	16	17	8	20	4	10	2	5	1
15	18	17	2	7	13	11	4	14	3	22	8	5	6	21	16	10	12	19	9	20	1
16	3	2	9	6	4	18	12	8	13	1	16	3	2	9	6	4	18	12	8	13	1
17	13	14	8	21	12	20	18	7	4	22	6	10	9	15	2	11	3	5	16	19	1
18	2	13	4	3	8	6	16	12	9	1	18	2	13	4	3	8	6	16	12	9	1
19	16	5	3	11	2	15	9	10	6	22	4	7	18	20	12	21	8	14	13	17	1
20	9	19	12	10	16	21	6	5	8	22	3	14	4	11	13	7	2	17	18	15	1
21	4	15	16	14	18	10	3	17	12	22	2	19	8	7	9	5	13	20	6	11	1
22	1	22	1	22	1	22	1	22	1	22	1	22	1	22	1	22	1	22	1	22	1

LCGs were popular for their simplicity, efficiency, and well-known theoretical properties. However, LCGs are not recommended because they have relatively short periods by current standards, and they lack equidistribution in dimensions higher than 1. Furthermore, LCGs have questionable empirical performances, and all LCGs have failed stringent empirical tests in L'Ecuyer and Simard (2007).

There are several proposed methods to improve the performance of LCGs. For example, Marsaglia Marsagila (1997) proposed the "multiply-with-carry" (MWC) generator given by

$$X_i = (BX_{i-1} + A_{i-1}) \bmod m, \tag{2.4}$$

with $A_i = \lfloor \frac{(BX_{i-1} + A_{i-1})}{m} \rfloor$ to be used as the "carry" for the next iteration and the initial A_{-1} set by the user. This class of generators is similar in form to Eq. 2.2 with "carry", A_i, as feedback for the next iteration. MWC was originally proposed as a generalization of the "add-with-carry" generator of Marsaglia and Zaman (1991). For efficiency considerations, it is recommended that $m = 2^{32}$. Additional theoretical studies on the MWCs can be found in Couture and L'Ecuyer (1997), Goresky and Klapper (2003). Compared to LCGs, MWC generators have much longer periods and better distribution properties, and they are reasonably efficient to compute. We will consider other methods to improve LCGs in Sect. 2.2.

2.2   Two Extensions of LCG

LCGs are very efficient. However, they have poor empirical performance, short periods, and inadequate uniformity in dimensions higher than one. These properties and others have earned the LCG a reputation as an unsuitable generator for modern, large-scale simulation tasks.

To find better classes of pseudo-random number generators, it is natural to consider extensions of the LCG. The first is the Multiple Recursive Generator (MRG) which generates the next pseudo-random number based on a kth order equation instead of the first order equation. MRGs indeed have better empirical performances, longer periods, and higher dimensions of uniformity. But they come at two costs: efficiency and finding the right parameters to meet the maximum period length, especially as the order k or modulus p increases. Next, we review the basics of MRGs, considerations when searching for MRGs, and special designs to improve efficiency.

The second extension of the LCG is the Matrix Congruential Generator (MCG), which extends the LCG to k-dimensions. Similar to the manner in which MRGs improve on LCGs, MCGs also benefit from better empirical performance, longer periods, and higher dimensions of uniformity. However, MCGs also have costs similar to those of MRGs. Later, MCGs and their mathematical properties are briefly discussed.

Extending the LCG with the proper design could prove to be the best balance of better empirical and theoretical performance while not costing too much in terms of efficiency or in searching for parameters that yield maximum period length for a given order k and modulus p.

2.2.1 Multiple Recursive Generator (MRG)

Multiple Recursive Generator (MRG) is the k-th order extension of the LCG. The kth order linear recurrence for an MRG can be defined as

$$X_i = \alpha_1 X_{i-1} + \alpha_2 X_{i-2} + \cdots + \alpha_k X_{i-k} \bmod p, \quad i \geq 0, \tag{2.5}$$

where $\alpha_1, \alpha_2, \ldots, \alpha_k$ are integers in $\mathbb{Z}_p = \{0, 1, \ldots, p-1\}$, $\alpha_k \neq 0$, and we can choose any k not-all-zero values as starting seeds, $(X_{-k}, X_{-(k-1)}, \ldots, X_{-1}) \neq (0, 0, \ldots, 0)$. When the order $k = 1$, the MRG reduces to a LCG.

A very well-known RNG is the **Linear Feedback Shift Register** (LFSR). The LFSR can be considered as a special case of MRG with $p = 2$, i.e., it works with binary bits. It is well-known that the maximum period of the MRG is $p^k - 1$ which is much larger than the period length of LCG ($p - 1$). Checking whether an MRG has maximum period length is equivalent to checking whether its *characteristic polynomial*

$$f(x) = x^k - \alpha_1 x^{k-1} - \alpha_2 x^{k-2} - \cdots - \alpha_k \tag{2.6}$$

is a k-degree primitive polynomial over $\mathbb{Z}_p$. See, for example, Knuth (1998). Note that there is a simple relationship between an MRG's *companion matrix* (defined below) and its characteristic polynomial. For an MRG defined in Eq. 2.5, we can define its corresponding companion matrix

$$\mathbf{M}_f = \begin{pmatrix} 0 & 1 & 0 & \ldots & 0 \\ 0 & 0 & 1 & \ldots & 0 \\ \vdots & \vdots & \vdots & \ddots & \vdots \\ 0 & 0 & 0 & \ldots & 1 \\ \alpha_k & \alpha_{k-1} & \alpha_{k-2} & \ldots & \alpha_1 \end{pmatrix}, \tag{2.7}$$

and it is straightforward to see that

$$f(x) = \det(x\mathbf{I} - \mathbf{M}_f) \bmod p. \tag{2.8}$$

For simplicity, we let MRG(k, p) denote the class of maximum-period k-th order MRGs with prime modulus p.

2.2.2 Matrix Congruential Generator (MCG)

The matrix congruential generator (MCG) is a natural k-dimensional extension of the LCG, defined by the recurrence relation

$$\mathbf{X}_i = \mathbf{B}\mathbf{X}_{i-1} \bmod p, \quad i \geq 0, \tag{2.9}$$

where $\mathbf{X}_i$ is a k-dimensional vector in $\mathbb{Z}_p^k$, $\mathbf{X}_{-1}$ is a non-zero vector, and the multiplier matrix $\mathbf{B}$ is a $k \times k$ matrix in $\mathbb{Z}_p^{k \times k}$. The characteristic polynomial $f_\mathbf{B}(x)$ of an MCG is defined as $\det(x\mathbf{I} - \mathbf{B}) \bmod p$. An MCG has a maximum period of $p^k - 1$ if and only if the matrix $\mathbf{B}$ has the order of $p^k - 1$, which is equivalent to its characteristic polynomial being primitive. Researchers such as Franklin (1964), Grothe (1987), L'Ecuyer (1990), Niederreiter (1986) have considered the MCG, with its maximum period determined by the order of the matrix. Different methods, including those proposed by Deng et al. (1992), Grothe (1987), are available for finding the matrix multiplier $\mathbf{B}$ with the maximum period in the context of MCGs.

The characteristic polynomial of an MCG is defined as

$$f_\mathbf{B}(x) = \det(x\mathbf{I} - \mathbf{B}) \bmod p. \tag{2.10}$$

An integer matrix $\mathbf{B} \in \mathbb{Z}_p^{k \times k}$ has order n if

$$n = \min_{j>0} \left\{ j : \mathbf{B}^j \bmod p = \mathbf{I} \right\}.$$

As a vector sequence $(\mathbf{X}_i)_{i \geq 0}$, an MCG has the maximum period of $p^k - 1$ *if and only if* the matrix $\mathbf{B}$ has the order of $p^k - 1$. It is well known that $\mathbf{B}$ has the order of $p^k - 1$ *if and only if* its corresponding characteristic polynomial $f_\mathbf{B}(x)$ defined in Eq. 2.10 is a primitive polynomial. In particular, for every generator in MRG(k,p) has a primitive characteristic polynomial, say, $f(x)$, the corresponding companion matrix $\mathbf{M}_f$ in Eq. 2.7 has the order of $p^k - 1$.

2.3   Linear-Feedback Shift Register (LFSR)

A special class of general RNGs with modulus, $p = 2$ and f in Eq. 2.1 a linear function is the linear feedback shift register (LFSR). The coefficients (α_j) for the LFSRs are either $\alpha_j = 0$ or 1 for $1 \leq j \leq k - 1$ with $\alpha_k = 1$ and an initial nonzero binary vector, $(X_{-k}, \ldots, X_{-1})$. The LFSR will achieve the maximum period $2^k - 1$ if the necessary and sufficient condition that the polynomial

$$f(x) = x^k - \alpha_1 x^{k-1} - \cdots - \alpha_k \tag{2.11}$$

is a primitive polynomial over $\mathbb{Z}_2$ is met (Knuth 1998). A computationally efficient algorithm that produces binary sequences that achieve the maximum period of $2^k - 1$ is given using a primitive polynomial with only three terms as

$$f(x) = x^k - x^{k-t} - 1, \quad 1 \le t \le k - 1. \tag{2.12}$$

and using a k-th order linear recurrence equation

$$X_i = (X_{i-k} + X_{i-t}) \bmod 2. \tag{2.13}$$

Watson (1962) tabulated some primitive polynomials with degree up to 100 and Knuth (1998) gives references for more primitive polynomials with modulus 2.

2.4   Generalized Feedback Shift Register (GFSR)

The Generalized Feedback Shift Register (GFSR) Pseudo-random Number Algorithm by Lewis and Payne (1973) is a method for generating pseudo-random numbers. It is based on the concept of a feedback shift register, which is a digital circuit that can be used to generate sequences of binary numbers with properties that appear random. The key idea is to use bitwise operations such as an exclusive-or ($\oplus$) operation. The basic formula for updating the state of the shift register at each iteration is as follows:

$$X_i = X_{i-k} \oplus X_{i-t} \tag{2.14}$$

This algorithm has played a pivotal role in the advancement of Very-Large Period (VLP) pseudo-random number generation. Notably, it has served as the foundation for the creation of other pseudo-random number generators, such as Matsumoto's Mersenne Twister (MT19937). This innovative approach to VLP generation has led to the development of more easily implementable VLPs using similar principles.

An efficient computer program for faster initialization of GFSR's was developed in Kirkpatrick and Stoll (1981). Because of their generating efficiency, the following generators became very popular in the field of computational physics:

$$R250: x_n = x_{n-250} \oplus x_{n-103}; \quad \text{and R1279:} y_n = y_{n-1279} \oplus y_{n-1063}.$$

Here, x_n and y_n are 32-bit integers that are computed iteratively from previous numbers generated. The numbers of initial seeds required are 250 and 1279 for R250 and R1279, respectively. It is common to produce the required seeds by another random number generator such as LCG(16807;$2^{31} - 1$).

The period lengths for R250 and R1279 are $2^{250} - 1 (\approx 10^{75.3})$ and $2^{1279} - 1 (\approx 10^{385.1})$, respectively. While R250 and R1279 are efficient with a long period length, their empirical performances had some problems as reported by Ferrenberg et al. (1992).

2.5 Additive Generator

It is inefficient for an LFSR to produce a random variate of 32 bits because it produces only 1 bit at each iteration. Alternatively, one can consider an additive generator of order k defined as follows:

$$X_i = (X_{i-t} + X_{i-k}) \bmod m \qquad (2.15)$$

with modulus m usually chosen as 2^e and $e = 32$ for generating efficiency. According to Knuth (1998, p. 28), when $m = 2^e$, with an appropriate choice of t in Eq. 2.12), the maximum period length of the additive generator is $2^{e-1}(2^k - 1)$. This period length is far less than $m^k - 1$ when m is a large prime number. Also, there is no equidistribution property for the additive generators. On the positive side, the main advantage of the additive generator is its generating efficiency. We denote the class of maximum-period additive generators of order k by ADD-k.

To consider a variation of GFSR, it is common to replace exclusive-or ($\oplus$) operation with simply $+$ or $-$ and modulus m operation.

This generator is denoted as LFG$(p, q;m)$. LFG is fast because no multiplication operation is required. However, it has several drawbacks: (1) it has a bad lattice structure because of its small coefficients (2) its empirical performance is generally poor, and (3) it is hard to find p, q, and m to maximize the period length of the generator, such as an additive lagged Fibonacci generator discussed in Marsaglia (1985). The equation for an additive lagged Fibonacci generator is

$$X_i = X_{i-k} \pm X_{i-t} \bmod 2^w. \qquad (2.16)$$

The maximum attainable period for these generators is $(2^k - 1)2^{w-1}$. For small values of k lagged Fibonacci generators may have poor statistical properties (Coddington 1994), while increasing k improves their performance but increases memory requirements.

Specifically, R software includes "Knuth-TAOCP" and "Knuth-TAOCP-2002", a more recent variant with different initialization, as user-selectable RNGs. They are lagged Fibonacci generators considered in Knuth (1998) as

$$X_i = X_{i-100} - X_{i-37} \bmod 2^{30}. \qquad (2.17)$$

2.6 MT19937

Matsumoto and Kurita (1992) proposed a twisted version of a GFSR. This is accomplished by multiplying a variate in the generator by a suitably chosen matrix. Perhaps the most popular RNG in current use for applications in computer simulation, the Mersenne Twister, proposed by Matsumoto and Nishimura (1998), is an example of a twisted GFSR (Gentle 2003). The MT19937 generator, based on the Mersenne prime $2^{19937} - 1$, is the most widely

implemented version of the Mersenne Twister algorithm and is the most widely used RNG
for computer simulations (Marsland 2011).

The MT19937 uses a matrix linear recurrence (with more than 100 non-zero terms) of
order 19937 over the binary finite field, $\mathbb{Z}_2$, and is related to matrix congruential generators
(Gentle 2003) to be discussed in the next section. MT19937 can provide fast generation of
32-bit, pseudo-random numbers, has a period length of $2^{19937} - 1 \approx 10^{6001}$, and equidistribution up to 623 dimensions. MT19937 is the default generator for many popular software
systems including Microsoft Excel, Python, standard C++, and R. It is also one of the RNGs
used in SPSS and in SAS.

Although MT19937 is currently the most popular RNG used for its long period length and
high-dimensional equidistributional property, MT19937 consistently fails linear complexity
tests in the TestU01 test suite (see, L'Ecuyer and Simard 2007).

2.7   Global Properties of the Generators

The introduction provides an overview of the theoretical properties of Multiplicative Congruential Generators (MCGs) and Multiply-With-Carry Generators (MRGs) as extensions
of Linear Congruential Generators (LCGs). MRGs involve a recurrence relation considering
multiple terms, while MCGs generalize MRGs by producing vector-valued variables. The
theoretical properties of MRGs include strong statistical justification, a large period length,
and equidistribution properties. Larger order values for a given prime modulus enhance these
advantages, but finding optimal parameters becomes challenging. The section on MCGs discusses the vector sequence generated by them, emphasizing the requirement for a maximum
period MCG to exhibit uniformity over the k-dimensional cube of Z_p^k. The column-wise output method is suitable for certain processors, producing high-dimensional random vectors.
The Cayley-Hamilton Theorem is used to link MCGs to MRGs, showing that a maximum-
period MCG can be viewed as running k copies of the same MRG with different starting
seeds, allowing for parallel simulation. The importance of choosing an efficient matrix multiplier for MCGs is highlighted, with references to efficient MCGs proposed in the literature.
Further detailed discussions on MCGs are provided in Chap. 10.

2.7.1   Period Length

As defined in Definition 1.3, the period length of a random number generator refers to the
number of values the generator can produce before it begins to repeat itself. In simpler terms,
it's the length of the sequence of random numbers before it cycles back to the beginning.

Having a longer period length is desirable in random number generators because it means
that the generator can produce a larger variety of random numbers before it repeats, which is
crucial for many applications where unique randomness is required over extended periods.

Period length is an important criterion for evaluating the quality and effectiveness of random number generators, especially in fields like cryptography, simulation, and gaming, where the randomness of the generated numbers is crucial.

2.7.2 Equidistribution

Equidistribution, also known as uniformity or evenness, refers to the property of a random number generator where each possible outcome has an equal probability of occurring. In other words, in an equidistributed sequence of random numbers, every value in the range has an equal chance of being selected.

Definition 21 (*Equidistribution in t dimensions*)

Suppose a RNG generates the sequence $X_{-k}, \ldots, X_{-1}, X_0, X_1, \ldots$ where $X_i \in \mathbb{Z}_p$, and let its period length be r. The RNG is said to be equidistributed in t dimensions, if the following property holds true for every $1 \leq d \leq t$: Consider all r possible distinct d-tuples of successive elements from i, i.e., $(X_i, \ldots, X_{i+d-1}), (X_{i+1}, \ldots, X_{i+d}), \ldots, (X_{i+r-1}, \ldots, X_{i+r+d-2})$. Then, amongst them all the possible p^t tuples will occur with almost equal frequency.

For many of the linear generators we consider, one can choose appropriate seeds and function, f such that all the p^t except possibly one will occur with equal frequency. If the X_i are indeed uniformly distributed in $\mathbb{Z}_p$ and are independent, then in expectation each such tuple will occur will equal frequency and thus the equidistribution property is defined to model this.

Example 2.4 (*Equidistribution property*)

Consider the generator $X_{-4} = 1, X_{-3} = 1, X_{-2} = 1, X_{-1} = 1, X_0, X_1, X_2, \ldots$, where $X_i = X_{i-4} + X_{i-1} \mod 2$ for $i \geq 0$. Here, the starting seeds are given by $X_{-4}, X_{-3}, X_{-2}, X_{-1}$ and all are 1. Here one is using the prime $p = 2$. One can confirm that the maximum period of $2^k - 1 = 15$ is indeed obtained. The following table shows the sequence, where one can imagine a sliding window over 4 successive members of the sequence that are always denoted as X_1, X_2, X_3, X_4 and the next element as X_5. One can check here that the equidistribution property is valid up to 4 dimensions. For example, for $t = 2$ dimensions, and the string tuple $(0, 1)$ it will occur in the table below at indexes, $5, 7, 11, 15$ and thus at $p^{k-t} = 2^{4-2} = 4$ times. Or, for example the tuple $(1, 0, 1)$ occurs at indexes $4, 6$. To make it easier to find where the tuples occur, the various rows are sorted by X_1, X_2, X_3, X_4 so that it makes it easier to find how many times they occur. All zero tuples, for example $(0, 0)$ or $(0, 0, 0)$, always occur one less time than the other tuples of the same size (Table 2.3).

Table 2.3 MRG sequences based on $X_i = X_{i-4} + X_{i-1}$ mod 2. The sorted sequence is sorted by X_1, X_2, X_3, X_4

index	Original sequence					Sorted sequence				
	X_1	X_2	X_3	X_4	X_5	X_1	X_2	X_3	X_4	X_5
1	1	1	1	1	0	0	0	0	1	1
2	1	1	1	0	1	0	0	1	0	0
3	1	1	0	1	0	0	0	1	1	1
4	1	0	1	0	1	0	1	0	0	0
5	0	1	0	1	1	0	1	0	1	1
6	1	0	1	1	0	0	1	1	0	0
7	0	1	1	0	0	0	1	1	1	1
8	1	1	0	0	1	1	0	0	0	1
9	1	0	0	1	0	1	0	0	1	0
10	0	0	1	0	0	1	0	1	0	1
11	0	1	0	0	0	1	0	1	1	0
12	1	0	0	0	1	1	1	0	0	1
13	0	0	0	1	1	1	1	0	1	0
14	0	0	1	1	1	1	1	1	0	1
15	0	1	1	1	1	1	1	1	1	0
16	1	1	1	1	0	Repeat				
17	1	1	1	0	1	Repeat				
	...					...				

More details will be discussed in Sect. 6.1.

2.7.3   Spectral Test

The spectral test is a theoretical test that provides a measure of uniformity of a MRG in dimensions beyond the order k of the MRG. It is often used to rank MRGs of the same order. A maximum period MRG of order k has a known period length. Assuming such an MRG generates numbers in $\mathbb{Z}_p$, it has order p^k. Moreover, it has the equidistribution property for dimensions $t \leq k$, see Definition 21. However, the equidistribution property does not hold for dimensions $t > k$. This means that if we consider the successive t tuples $(X_i, X_{i+1}, \ldots, X_{i+t-1})$ over the entire sequence, all the possible p^t tuples are not generated. If we consider these tuples in the t dimensional space $\mathbb{Z}_p^t$, then not all possible tuples are present. One way to measure how uniformly distributed the generated tuples are is to consider how they are covered by parallel hyperplanes with equal gaps. Different parallel families achieve different gaps, but we focus on one that has the maximum gap possible, as these are the ones that "expose" the non-uniformity best. Intuitively, if one were "looking" in the normal direction of that family of hyperplanes, the generated tuples would appear to have the maximum gaps. The spectral test computes this gap and uses that to measure the quality of the MRG. The spectral test can be computed given the MRG definition. A more detailed discussion appears in Chaps. 6 and 7.

2.7.4   Statistical Justification

Random number generators (RNGs) are fundamental tools in various fields, from cryptography to simulations in scientific research. The statistical justification for their use hinges on several critical factors:

1. Uniformity: The primary requirement for an RNG is that the numbers it produces should be uniformly distributed over the desired range. This means each number within the range should have an equal probability of being selected. Statistically, uniform distribution ensures that the generated numbers are unbiased and representative of the entire range.
2. Independence: Another key statistical property is the independence of successive numbers generated. In an ideal RNG, the generation of one number should not influence the generation of the next. This lack of correlation ensures that the numbers do not follow any discernible pattern, making them suitable for applications requiring randomness.

More advanced statistical tests for evaluating the randomness of a sequence will be discussed in Sect. 2.8. The theoretical support will be provided in Sect. 5.4

2.8   Local Empirical Evaluation of the Generators

Statistical tests for randomness are procedures used to assess whether a sequence exhibits characteristics consistent with randomness or if there are discernible patterns, structures, or deviations from randomness. The fundamental idea is to subject the sequence to a set of statistical analyses to determine whether it behaves as one would expect from a random process. In this context of randomness, it means that values in the sequence are independent and identically distributed (i.i.d.), with no predictable order or pattern. Randomness is a crucial assumption in various statistical analyses, and ensuring that the sequence is truly random is important for making valid inferences.

2.8.1   Diehard Tests

The Diehard test suite is a set of statistical tests for measuring the quality of a random number generator. It was developed by Marsaglia (1996). It contains originally 15 statistical tests. It is later extended to the Dieharder tests and the *TestU01* library. The details of *TestU01* will be presented in Sect. 2.8.3.

2.8.2   NIST Statistical Test Suite

The NIST Statistical Test Suite is a set of tests provided by the National Institute of Standards and Technology (NIST) to assess the randomness and quality of a random number generator. The NIST Statistical Test Suite is part of the NIST Special Publication 800-22, titled "A Statistical Test Suite for Random and Pseudo-random Number Generators for Cryptographic Applications" (last known update is 2014). It contains 15 tests:

1. Monobit Frequency,
2. Block Frequency,
3. Run,
4. Longest Run,
5. Binary Matrix Rank,
6. Discrete Fourier Transform,
7. Non-overlapping Template Matching,
8. Overlapping Template Matching,
9. Maurer's "Universal statistical" test,
10. Linear Complexity,
11. Serial,
12. Approximate Entropy,
13. Cumulative Sums,
14. Random Excursions, and
15. Random Excursions Variant.

More details can be found on the website https://csrc.nist.gov/projects/random-bit-generation/documentation-and-software. Given that the NIST tests overlap within the *TestU01* library, and considering the widespread usage of the *TestU01* as a test suite that is actively maintained and updated (with the latest updates as of July 2023 available at https://github.com/umontreal-simul/TestU01-2009/), we will delve into a discussion of these tests in Sect. 2.8.3.

2.8.3 TestU01

Overview of TestU01 Library

While random number generators (RNGs) are used for a variety of purposes, including simulation and cryptography, it is important to make sure that one is using an RNG that is empirically sound. This means that the sequence of generated variates closely resembles a uniform distribution and is hard to distinguish from a truly random sequence. The *TestU01* library allows RNGs to undergo a number of statistical tests for uniformity. The *TestU01* library differs from other testing packages, such as DIEHARDer, the NIST package, and others, in that *TestU01* is more stringent, contains more statistical tests, and has more flexibility for testing generators compared to the other packages. It also allows for one to test both random numbers as well as random bits from a generator. This overview will mainly focus on the former, but implementation of the latter can easily be done with a similar approach and is theoretically related.

It is important to note that the *TestU01* library does not contain every imaginable statistical test, as there are an infinite number of possible tests that could be created. The library does contain a variety of different tests which gives a good idea of the quality of the RNG. These tests are also not meant to prove that an RNG will be "perfect", but passing the tests gives confidence that the RNG is virtually indistinguishable from a truly uniform sequence. "The difference between the good and bad RNGs, in a nutshell, is that the bad ones fail very simple tests whereas the good ones fail only very complicated tests that are hard to figure out or impractical to run" (L'Ecuyer and Simard).

Reading of the p-values

The *p-value* is a statistical measure that helps assess the evidence against a null hypothesis in a hypothesis test. In the context of RNG tests, the *p-value* indicates the probability of observing sequences as extreme as the ones obtained, assuming that the null hypothesis is true. When evaluating RNGs, statistical tests involve the computation of *p-values*. While evaluating the sequences generated by an RNG, the collection of the *p-values* should follow a uniform distribution. Thus the collection of *p-values* can be used to measure the uniformity of the generator.

The *TestU01* library reports a test of an RNG when the *p-values* are extremely close to 0 or 1 (the values outside $[0.001, 0.999]$). Since there are finite tests in the TESTU01 library (10 tests on Small Crush, 96 tests on Crush, and 106 tests on Big Crush), it is unlikely to obtain a *p-value* less than 10^{-10} or greater than $1 - 10^{-10}$. The package will reject an RNG if one *p-value* is less than 10^{-10} or greater than $1 - 10^{-10}$. In other words, an RNG anticipated to successfully pass a test with k% probability is likely to be flagged as "suspicious" approximately k% of the time. In instances where generators unmistakably fail tests, the *p-values* swiftly approach either 0 or 1.

In testing the RNGs, the *TestU01* library utilizes what is known as a *two-level* procedure, which essentially means that a test statistic (Y_i) is generated independently N times, and then transformed so that the transformations (U_i) are i.i.d. uniform. This is beneficial in allowing the test to have a larger sample size which increases its power, and it also allows for testing at the local level instead of the global level. The latter benefit could be done through a spacing transformation, which is done by seeing how far apart successive $U_{(i)}$ are from each other. This would help detect clustering of an RNG and help determine if it is a good generator or not. Additionally, it should be noted that most test statistics in the *TestU01* library will follow a chi-square, normal, or Poisson distribution.

In addition to the test mentioned above, the library also carries out tests on a single stream of numbers by measuring global uniformity and clustering, as well as implementing run and gap tests. When testing sub-sequences, a common technique is to divide the unit hyper-cube into partitions and count the number of hits per piece, or measure how long it takes a partition to reach a certain number of hits. Likewise, once the hyper-cube is partitioned, one can measure the time gaps between visits to the cell. Additional types of tests, such as birthday spacing, test for randomness. For more information consult L'Ecuyer and Simard (2007) for an in-depth explanation of the theory behind the tests.

Background Information about Statistical Tests

Chi-square test

One popular test for testing the goodness-of-fit of a sequence of data is the chi-square test, first proposed by Pearson (1900). If we have data which falls into a finite number of categories, say k categories, then we might look into using the Chi-square test. Assume we have n independent observations, and that these observations fall into one of k categories. Then, E_i is the theoretical frequency for the observation that will fall into category i, and O_i is the observed frequency for the number of observations that do fall into category i. Then the test statistic will be

$$T = \sum_{x=1}^{k} \frac{(O_i - E_i)^2}{E_i}.$$

We can then consult tables or software to determine the p-value of the test statistic T and then make a conclusion on whether the sample data fit the desired frequency table well.

Kolmogorov-Smirnov test

If our observations fall in a range with infinitely many values, say $U(0, 1)$, then we might look into the Kolmogorov-Smirnov (KS) test (Massey 1951). This will allow us to compare the difference between theoretical distribution $F(x)$ and our empirical distribution $F_n(x)$. We should note that if we have n independent observations ordered from smallest to largest, $X_{(1)}, X_{(2)}, \ldots, X_{(n)}$, then the empirical distribution $F_n(x)$ will be

$$F_n(x) = \frac{\text{number of } X_{(1)}, X_{(2)}, \ldots, X_{(n)} \text{that are } \leq x}{n}.$$

The test statistic will then be

$$K = \sup_x |F(x) - F_n(x)|.$$

We can then consult tables or software to determine the p-value of the test statistic K and then make a conclusion on whether the sample data fit the desired distribution well.

Statistical Tests

While there are a great number of statistical tests currently available in the literature, this section aims to discuss a few types of commonly used tests, as described in Knuth (1998). For the following section, we will denote $\langle U_n \rangle = U_0, U_1, U_2, \ldots$ as a sequence of numbers between 0 and 1. Additionally, we will denote $\langle Y_n \rangle = Y_0, Y_1, Y_2, \ldots$ as a sequence of numbers between 0 and $d - 1$. Because of this, we could write $Y_i = \lfloor dU_i \rfloor$. There are a number of different additional ways that we could transform $\langle Y_n \rangle$ into the sequence $\langle U_n \rangle$.

Frequency (Equidistribution) Test

One of the first tests we might think of is to test if the numbers do in fact follow a uniform distribution. Given the sequence $\langle U_n \rangle$, we could carry out a Kolmogorov-Smirnov test with $F(x) = x$ for $0 \leq x \leq 1$. Additionally, we could use the chi-square test to determine if the sequence, $\langle Y_n \rangle$, is integer-valued or broken up into a finite number of categories.

Serial Test

In addition to wanting the sequence to be independent and uniform, we also wish to have successive pairs (and t-tuples) to be independent and uniform. The Serial test, proposed by Good (1953), is used to see if successive pairs are independent and uniformly distributed. The general idea of this test is to run the frequency test for higher dimensions. Say we want to look at successive pairs, then we will partition the two-dimensional $(0, 1)$ square into $k = d^2$ cells of area $1/k$. We will then count the number of times the pair (Y_{2i}, Y_{2i+1}) falls in each category, for $0 \leq j < n$. The chi-square test is then run to compare the expected values for each square vs the observed values. This test can be run for higher dimensions (successive t-tuples) with the same idea of dividing the hyper-cube $(0, 1)^t$ into $k = d^t$ cubes.

Gap Test

It may be useful to determine the length of time (or "gap") between a sequence's visit to a certain range. Given $0 \leq \alpha < \beta \leq 1$, we want to find the lengths of the sub-sequence $U_j, U_{j+1}, \ldots U_{j+r}$ such that both U_j and U_{j+r} are in the interval (α, β) but the other values do not fall in the interval. We will run the process n times (to end up with n gaps) and count the number of times each gap length occurs. Since we are dealing with the interval (α, β), we can denote the probability of the next number being in the interval as $p = \alpha - \beta$. We will then compare our observed gap lengths with the expected gap lengths (following a geometric distribution) using a chi-square test.

Poker Test

Similar to a poker hand, we can also run a test called the Poker test Kendall and Babington-Smith (1939) to count the number of times each type of "hand" occurs. In this test, we will take n sets of five consecutive integers $Y_{5j}, Y_{5j+1}, \ldots Y_{5j+4}$, for $0 \leq j < n$ and count the number of distinct values in the set of five. Due to possibly having the range of Y_i be very large, we might need to partition the domain into categories or take $Y_i = \lfloor dU_i \rfloor$ where d is some chosen value. With the 5 consecutive integers, we can either have 5 distinct values, 4 distinct values, $\ldots$, or 1 distinct value. After determining the categories for the n sets, we can then do a chi-square test to compare the observed probabilities with the predicted probabilities. Knuth (1998) notes the predicted probabilities of k-tuples for r distinct values is

$$p_r = \frac{d(d-1)\ldots(d-r+1)}{d^k} \begin{Bmatrix} k \\ r \end{Bmatrix},$$

where $\begin{Bmatrix} k \\ r \end{Bmatrix}$ are Stirling numbers. We can then compare the test statistic to determine the p-value.

Coupon Collector's Test

The Coupon Collector's Test (Greenwood 1955) examines how long it will take to "collect" all of the coupons. Given the sequence $Y_j, Y_{j+1}, \ldots$, where $Y_i = \lfloor dU_i \rfloor$ for some d value, we calculate the length of segments needed to "complete" the set of integers from 0 to $d-1$. We repeat this process n times with different sequences and compare the observed lengths to the predicted lengths using a chi-square test. Derivations of the predicted probabilities can be found in Knuth (1998) as well as Von Schelling (1954).

Permutation Test

Another test for a sequence of data is the Permutation test. This test divides the sequence into n sub-sequences of t elements each, giving us $U_{jt}, U_{jt+1}, \ldots U_{jt+t-1}$ for $0 \leq j < n$. We then classify each sub-sequence according to the relative ordering of the numbers. Note that we will have $t!$ possible relative orderings. We count the number of times each ordering occurs and run a chi-square test comparing the observed frequency versus the expected probability of $1/t!$.

Run Test

The Run test focuses on the "runs up" and "runs down" a sequence may exhibit. That is, the test looks at the length of the monotone sections of the sequence. This test is different than previous ones, as the chi-square test cannot be carried out on the counts of the runs, as they are not independent of each other. This is because a long run will normally be followed by a few short runs. Due to not having the independence condition met, we will need to find another test statistic. Details about how to carry out this test are presented in Knuth (1998).

Maximum-of-t Test

For the Maximum-of-t test, we take a sub-sequence of length t and record the maximum value in the sequence a total of n times. Let $V_j = \max(U_{tj}, U_{tj+1}, U_{tj+t-1})$ for $0 \leq j < n$. We then apply the Kolmogorov-Smirnov (KS) test on the sequence of $\langle V_i \rangle$ with the distribution function $F(x) = x^t$ for $0 \leq x \leq 1$. We choose this distribution function as the sequences are independent, so the probability $P(U_i < x) = x$ and thus the probability $P(U_1, U_2, \ldots U_t) = x^t$ as $xx \cdots x = x^t$.

Collision Test

The chi-square test should only be used when a nontrivial number of items are expected for each category. When the number of categories is greater than the number of observations then the criteria would not be met. The collision test is rooted in the idea that we have m urns and n balls with $n \ll m$. If we were to randomly throw the balls into the urns then most balls will land in urns that are empty, but if more than one ball lands in the same urn then we have a "collision". We then count the number of collisions that occur. The generator will pass the Collision test if there aren't too few or too many collisions. The probability of c collisions is then

$$\frac{m(m-1)\ldots(m-n+c+1)}{m^n} \left\{ \begin{matrix} n \\ n-c \end{matrix} \right\},$$

where $\left\{ \begin{matrix} n \\ n-c \end{matrix} \right\}$ are Stirling numbers. We can then check the p-value to make sure it is not too low or too high.

Birthday Spacings Test

The Birthday Spacings test proposed by Marsaglia (1985) is similar to the Collision test in a few ways. Instead of m urns and n balls, we will think about m days and n birthdays. Given the n birthdays as $Y_1, Y_2, \ldots Y_n$, with $0 \leq Y_i < m$, we then order the sequence as $Y_{(1)}, Y_{(2)}, \ldots Y_{(n)}$. We then calculate the spacing between the successive observations as $S_i = Y_{(i+1)} - Y_{(i)}$ and then we sort the spacings as $S_{(1)}, S_{(2)}, S_{(n)}$ and count the number of equal spacings called R. We then repeat this process many times, say N times, and carry out a chi-square test on the observed R's and the expected distribution. The expected distribution should follow a Poisson distribution with mean equal to $Nn^3/(4m)$. Additional explanation on the distribution can be found in L'Ecuyer and Simard (2007).

Serial Correlation Test

Correlation is often used to indicate whether two quantities might be relatively independent of each other. We can calculate the correlation of the observations $(U_0, U_1, \ldots U_{n-1})$ and $(U_1, U_2, \ldots U_n)$ with the Serial Correlation Test. The test statistic will be

$$C = \frac{n(U_0U_1 + U_1U_2 + \cdots U_{n-2}U_{n-1} + U_{n-1}U_0) - (U_0 + U_1 + \cdots U_{n-1})^2}{n(U_0^2 + U_1^2 + \cdots U_{n-1}^2) - (U_0 + U_1 + \cdots U_{n-1})^2}.$$

Since $U_0 U_1$ and $U_1 U_2$ are not independent of each other, we do not expect the serial corre-lation to be exactly 0. A "good" value of C should be between $\mu_n \pm 2\sigma_n$ where $\mu_n = \frac{-1}{n-1}$ and $\sigma_n = \frac{n^2}{(n-1)^2(n-1)}$ for $n \geq 2$

Organization of TESTU01

The library is organized into four different modules. The 'u' module is for implementing RNGs, the 's' module is for implementing statistical tests, the 'b' module is for implementing predefined batteries of tests, and the 'f' module is for implementing tests on a family of generators. Within the 'u' module, there are certain commands and objects that are required to define an RNG. The 'unif01_Gen' object contains the name, the parameters, the state, and a function ('GetU01' and/or 'GetBits') to generate the random variates/bits and advance the generator one step. The 'GetBits' function will return a block of 32 bits and the 'GetU01' will return a number between 0 and 1. There are also additional optional features that allow a generator to undergo an output filter (which drops a certain number of bits or combines successive values). Working with parallel generators and combination generators is also possible.

Individual tests can be run in the 's' module. In order to do this, one simply passes the generator object and required parameters through the testing function. For more information, consult (L'Ecuyer and Simard 2013), which is a user's guide for the *TestU01* library. Running individual tests may be beneficial if one is interested in designing/altering an RNG in order to pass a specific test.

Potentially the most useful module in testing an RNG is the 'b' module. This module contains several predefined batteries of statistical tests. There are three main batteries that we will concern ourselves with; *SmallCrush*, *Crush*, and *BigCrush*. The approximate run times for these batteries are about 14 s, 1 h, and 5.5 h respectively, with the number of tests in each battery being 10, 96, and 106 respectively. Therefore, when testing an RNG, it is recommended to start with SmallCrush, and if all tests are passed, work up to BigCrush. For the specific tests in each battery, consult the users-guide L'Ecuyer and Simard (2013). Similar batteries also exist for testing random bits.

Installing and implementing the *TestU01* library for the first time is relatively straight-forward and is done in the terminal. The source files can be downloaded http://simul.iro. umontreal.ca/testu01/tu01.html as a zipped file. After unzipping the file into your preferred folder/directory, you will then need to configure the source code into a different folder. Then a call of 'make' and 'make install' will create and format the *TestU01* library into the desired folder. Finally, after this is done and no errors are given, you will need to set the LD_LIBRARY_PATH, and LIBRARY_PATH, and the C_INCLUDE_PATH. All infor-mation is available in the README file contained in the ZIP file as well as in a Review of TESTU01 McCullough (2006). An example of how to install the *TestU01* library can be found below. Below, the system prompts and commands are common UNIX ones for the user whose home directory is at '/Users', with the file being unZipped in the 'TestU01' folder and the test suite being configured in the 'TEST-DIRECTORY' folder.

Design and Efficient Implementation of Large-Order Multiple Recursive Generators

Using computer-generated random numbers to simulate complex models or processes has become a common practice in scientific research. The quality of simulation studies heavily depends on the quality of the random number generators used. Since classical generators like Linear Congruential Generators (LCGs) proposed by Lehmer (1951) are known to have several defects (such as a relatively short period length by today's standard, questionable empirical performances, and lack of higher-dimension uniformity), Multiple Recursive Generators (MRGs) have replaced LCGs as one of the most popular generators.

MRGs of order k can achieve a maximal period of p^k, where p is the prime number used in the modular recurrence. A maximal period MRG also exhibits the property of equidistribution in spaces with dimensions up to k.

However, an MRG with a large order k needs a memory space of size k to store the k-states of the MRG. While this high memory requirement may seem like a shortcoming of such MRGs, memory cost has been drastically reduced in recent years and is now considered minimal even for large values of k, such as 30000. The one-time cost for the initialization of k states is also minimal, especially for large-scale simulation studies. Moreover, increasing the order k by 1 will increase the period length by p fold. However, if the memory space requirement or initialization cost is an issue for the application, one can consider using a smaller value of k.

In this chapter, we explore the central problem of finding such maximal period MRGs for use. In Sect. 3.2, we define a generic a generic way of forming efficient MRGs. The definition can be specialized to form different classes, discussed next. For example, we explore a set of portable and efficient large-order Multiple Recursive Generators (MRGs) characterized by identical coefficients for non-zero multipliers, as introduced by Deng and Xu (2003), known as the DX-k-s generators. We systematically identify and compile a category of

L. Deng et al., *Random Number Generators for Computer Simulation and Cyber Security*, Synthesis Lectures on Mathematics & Statistics, https://doi.org/10.1007/978-3-031-76722-7_3

DX generators with substantial orders, with the most extensive order being $k = 10007$ and a corresponding period length of approximately 10^{93384}. Additionally, we provide the outcomes of rigorous empirical tests conducted on these DX generators, demonstrating their successful passage of these assessments. Other classes are also discussed such as DL, DS, and DT classes.

3.1 Advantages and Challenges of Large-Order MRGs

Multiple recursive generators (MRGs) have become one of the most commonly used random number generators in computer simulation. MRGs are based on the k-th order linear recurrence

$$X_i = (\alpha_1 X_{i-1} + \cdots + \alpha_k X_{i-k}) \bmod p, \ i \geq k \tag{3.1}$$

for any initial seeds $(X_0, \ldots, X_{k-1})$ all in $\mathbb{Z}_p$ and not all of them being zero. Here the modulus p is a large prime number. To get variates $U_i \in [0, 1]$, X_i can be transformed using $U_i = X_i/p$. To avoid the possibility of obtaining 0 or 1, it is recommended $U_i = (X_i + 0.5)/p$ (see, e.g., Deng and Xu 2003).

3.2 General Class of Efficient Generators

The random numbers generated by any specific algorithm are "systematic" and therefore are neither truly independent nor purely random. A set of criteria is thus desirable for selecting /comparing random number generators, as briefly discussed in Section 2.7 and Section 2.8. It is common to consider five major issues for the comparison: (1) period length, (2) computing efficiency, (3) portability, (4) theoretical justification on randomness, and (5) empirical performance. For period length, L'Ecuyer (1997) showed that the set of all d ($> k$) tuples in the generated MRG sequence as given in Eq. 3.1 are distributed in equidistant parallel hyperplanes at distance $D > (1 + \sum_{i=1}^{k} \alpha_i^2)^{-1/2}$. Therefore, a large $\sum_{i=1}^{k} \alpha_i^2$ is a necessary (but not sufficient) condition for a good MRG. Consequently, it is better to have more nonzero terms with larger α_i^2 in general. While this may give a nice justification for considering MRGs with many nonzero terms, we need also to consider their generating efficiency.

To construct a general class of efficient generators, consider a special class of MRGs that have at most two different nonzero coefficients, say, A and B. Let S_A and S_B be two index sets defined as $S_A = \{j \mid \alpha_j = A\}$ and $S_B = \{j \mid \alpha_j = B\}$, such that $S_A \cap S_B = \emptyset$. Deng et al. (2008b) proposed a general class of generators of the following form:

$$X_i = A \sum_{j \in S_A} X_{i-j} + B \sum_{j \in S_B} X_{i-j} \bmod p. \tag{3.2}$$

A simple way to make this class of generators computationally efficient is to have only a few elements in both S_A and S_B. To design efficient large-order MRGs, Deng and Lin (2000) proposed a fast multiple recursive generator (FMRG). It is later expanded to the DX generator Deng and Xu (2003).

3.2.1   DX Generators

The DX generator, initially introduced by Deng and Xu (2003) and subsequently expanded upon by Deng (2005), constitutes a system of portable and efficient MRGs characterized by a modulus m and an order k. The efficiency of DX generators stems from the equal nature of all nonzero coefficients α_i in the recurrence. Notably, the DX generator encompasses the Fast MRG (FMRG) proposed by Deng and Lin (2000) as a specialized case. A key advantage of DX generators lies in the requirement of only a single multiplication for recurrence computation, resulting in faster execution compared to the general case.

When the order k is large, generating a general MRG can be slow. To address this, researchers have introduced special forms of MRGs with only a small number of nonzero terms. Deng (2005) introduced a class of DX-k-s-t generators:

1. DX-k-1-t $(\alpha_t = 1, \alpha_k = B)$, $1 \leq t < k$,

$$X_i = X_{i-t} + BX_{i-k} \bmod p, \quad i \geq k. \tag{3.3}$$

2. DX-k-2-t $(\alpha_t = \alpha_k = B)$, $1 \leq t < k$,

$$X_i = B(X_{i-t} + X_{i-k}) \bmod p, \quad i \geq k. \tag{3.4}$$

3. DX-k-3-t $(\alpha_t = 1 = \alpha_k = B)$, $1 \leq t < \lfloor k/2 \rfloor$,

$$X_i = B(X_{i-t} + X_{i-\lfloor k/2 \rfloor} + X_{i-k}) \bmod p, \quad i \geq k. \tag{3.5}$$

4. DX-k-4-t $(\alpha_t = 1 = 1 = \alpha_k = B)$, $1 \leq t < \lfloor k/3 \rfloor$,

$$X_i = B(X_{i-t} + X_{i-\lfloor k/3 \rfloor} + X_{i-\lfloor 2k/3 \rfloor} + X_{i-k}) \bmod p, \quad i \geq k. \tag{3.6}$$

The notation $\lfloor x \rfloor$ denotes the floor function of a number x, returning the smallest integer $\geq x$. In these DX-k-s-t generators, k represents the order, s specifies the number of nonzero terms with the equal coefficient B, and t indicates the distance of the first nonzero term in the recurrence equation. Given that multiplication is slower than addition or subtraction, the efficiency of DX generators is evident as they require only one multiplication and a small number of additions.

Example 3.1 (*DX-1597-4-t*)

For specific parameters $k = 1597$ and $m = 2^{31} - 1$, Deng (2005) identified several DX generators with a maximum period of 10^{14903}:

$$X_n = B(X_{n-t} + X_{n-533} + X_{n-1065} + X_{n-1597}) \bmod m, \quad n \ge 1597.$$

When $t = 1$, the multiplier B achieving the maximum period found are (a) $B = 1854$ (smallest), (b) $B = 44875$ ($B < \sqrt{m}$), (c) $B = 512675$ ($B < 2^{19}$), and (d) $B = 1073741362 = 29746 \times 36097$ ($B < 2^{30}$). See the discussion in Deng (2005) for its usefulness and a portable implementation for various B listed. Special forms of $B = 2^p \pm 2^q$ were also explored, enhancing the efficiency of DX generators, as proposed by Wu (1997) for LCG. For $t = 3$, two generators identified were (e) $B = 2^{29} + 2^8 = 536871168$ and (f) $B = 2^{28} + 2^{11} = 268437504$, denoted as DX-1597-(a) to DX-1597-(f).

While MRGs offer an increased maximum period of $p^k - 1$, their efficiency is inferior to LCG due to multiple involved multiplications. To enhance generation speed, Grube (1973), L'Ecuyer and Blouin (1988), L'Ecuyer et al. (1993) suggested employing only two nonzero terms α_j and α_k ($1 \le j < k$) of the MRG and provided portable implementations satisfying these conditions. Deng and Lin (2000) proposed setting many coefficients of α_i in an MRG to be 0 and/or ± 1, introducing a fast multiple recursive generator (FMRG) as a special MRG with maximum period, $p^k - 1$.

DX generators, like any PRNGs with very few nonzero coefficients, typically have the drawback of a "bad initialization effect". This effect is observed when the k-dimensional state vector is close to the zero vector and the subsequent numbers generated tend to stay within a neighborhood of zero for many generations before they can break away from this near-zero state, a property not desirable in the sense of randomness. As a result of bad initialization, two generated sequences using the same DX generator with nearly identical state vectors may not depart from each other quickly enough. This "bad initialization effect" was first observed by Panneton et al. (2006) for MT19937.

Classes of MRGs can be designed to overcome this "bad initialization effect" by including many non-zero terms, however, these MRGs tend to be inefficient. It is possible to find efficient implementations for certain generators by rewriting Eq. 3.2 as a simple higher-order recurrence equation (see Deng and Shiau 2015, for details). Here are a few examples of DX-k-s-t generators.

(a) For DX-k-1-t in Eq. 3.3, we choose $A = 1$, $S_A = \{t\}$, and $S_B = \{k\}$.
(b) For DX-k-2-t in Eq. 3.4, we let $S_A = \emptyset$ and $S_B = \{t, k\}$.
(c) For DX-k-3-t in Eq. 3.5, we let $S_A = \emptyset$ and $S_B = \{t, \lfloor k/2 \rfloor, k\}$.
(d) For DX-k-4-t in Eq. 3.6, we let $S_A = \emptyset$ and $S_B = \{t, \lfloor k/3 \rfloor, \lfloor 2k/3 \rfloor, k\}$.

Notice that the parameter s in the notation DX-k-s-t is equal to $\#(S_B)$, the number of indices in the set S_B.

When the numbers of the indices in S_A and/or S_B are large (i.e., $\#(S_A)$ and/or $\#(S_B)$ are large), it is still possible to find an efficient implementation for some generators as in Eq. 3.2. The idea is to impose a special structure on both S_A and S_B so that Eq. 3.2 can be rewritten as a simpler higher-order recurrence equation. Two classes of such generators are discussed next.

In summary, not all MRGs are efficient in generating random numbers or exhibit good empirical/theoretical performance. Therefore, when utilizing the algorithm GMP (will be presented in Sect. 4.2.1) to search for optimal maximum-period MRGs, it is crucial to confine the search within specific classes of MRGs. This study explores the extension of DX/DL/DS/DT generators in this context.

3.2.2   DL Generators

Examining the structure of generators of the form Eq. 3.2 with restrictions on the forms of S_A and S_B Deng et al. (2008b) propose a class of DL generators with $S_A = \{1, 2, \ldots, t-1\}$ and $S_B = \{t, t+1, \ldots, k\}$ for $1 \leq t < k$, and $g = 1$. Then

$$X_i = A(X_{i-1} + X_{i-2} + \cdots + X_{i-t+1}) + B(X_{i-t} + X_{i-t-1} + \cdots + X_{i-k}) \bmod p, \tag{3.7}$$

where t can be useful to expand the search space for maximal period DL generators. An efficient implementation of DL generators is discussed in Deng and Shiau (2015). Note that the t here plays a similar role as the t in DX-k-s-t. Using higher-order recurrence equation, DL generators can be implemented efficiently as:

$$X_i = X_{i-1} + A(X_{i-1} - X_{i-t}) + B(X_{i-t} - X_{i-(k+1)}) \bmod p, \quad i \geq k+1, \tag{3.8}$$

where $X_0, X_1, \ldots, X_{k-1}$ are the initial seeds and X_k is computed according to Eq. 3.7. Li (2005), Deng (2005) considered and tabulated this general class of DL generators for the order $k \leq 1709$.

From the above equation, we can see that only two multiplications and several additions/subtractions are needed to calculate the next value. To further improve the efficiency and the portability, we take the coefficient $A = 0, -1, 1$, or $-B$. Using these special values of A, we can reduce one multiplication and save some addition or subtraction operations. In particular, when $A = 0$, it leads to a simpler form

$$X_i = B(X_{i-t} + X_{i-t-1} + \cdots + X_{i-k}) \bmod p, \quad i \geq k, t \geq 1. \tag{3.9}$$

For simplicity, we consider this special case of $A = 0$ and $t = 1$ as a representative for the DL-k generators. Such DL generator can be implemented efficiently by:

$$X_i = X_{i-1} + B(X_{i-1} - X_{i-(k+1)}) \bmod p, \quad i \geq k + 1. \tag{3.10}$$

Marsaglia (1996) considered a special DL generator of small order ($k = 3$) with $B = 2^{10}$, $p = 2^{32} - 5$ and $B = 2^{20}$, $p = 2^{32} - 209$. Since the order $k = 3$ is small, there is no need to consider its higher order implementation.

3.2.3　DS Generators

We consider another variation of generators with few nonzero terms:

$$X_i = B \sum_{j=1}^{k} X_{i-j} - D X_{i-d} \bmod p. \tag{3.11}$$

By introducing parameters for the multipliers B, D, and index d, we can expand the search parameter space for the maximum period generators in that class. In addition, we can increase the complexity of the recurrence generating equation for the corresponding generators. We refer to this class of generators as DS generators. Like DL generators, DS generators can be efficiently implemented via $(k + 1)$ order of recurrence equations:

$$X_i = X_{i-1} + B(X_{i-1} - X_{i-k-1}) - D(X_{i-d} - X_{i-(d+1)}) \bmod p, \quad i \geq k + 1, \tag{3.12}$$

where $X_0, X_1, \ldots, X_{k-1}$ are the initial seeds and X_k is computed according to Eq. 3.11.

There are several interesting special cases for the DS generators. When $D = 0$, DS generator is reduced to a special case of DL generators with $t = 1$ as in Eq. 3.9. When $D = B$, the DS generator has exactly one zero term at the d-th term:

$$X_i = B \sum_{j=1, j \neq d}^{k} X_{i-j} \bmod p \tag{3.13}$$

which can be efficiently implemented as

$$X_i = X_{i-1} + B(X_{i-1} - X_{i-d} + X_{i-d-1} - X_{i-k-1}) \bmod p, \quad i \geq k + 1, \tag{3.14}$$

The parameter for the index d for the zero coefficient can be arbitrarily chosen. For simplicity, $d = \lfloor k/2 \rfloor$ for the DS-k generators considered is chosen.

Comparing Eqs. 3.10 and 3.14, we can see that higher order implementation of DS generators has a more complex recurrent structure than that of DL generators. Therefore, DS generators may have a better lattice structure for dimensions larger than k as will be described in Chap. 6. Both DS and DL have the "perfect" lattice structure for dimensions up to k. Like DL generators, DS generators can be efficiently implemented using a $(k+1)$st order recurrence equation (see Deng and Shiau 2015).

3.2.4 DT Generators

Another class of MRGs, called DT generators, with many nonzero terms and unequal weights on each term as discussed in Deng et al. (2009b) is given by

$$X_i = B^k X_{i-1} + B^{k-1} X_{i-2} + \cdots + B X_{i-k} \bmod p, \quad i \geq k. \qquad (3.15)$$

DT generators can also be efficiently implemented with a $(k+1)$-order recurrence equation (see Deng and Shiau 2015).

3.2.5 Motivations Behind Various Efficient Generators

With only very few non-zero terms in the MRG generating equation, FMRG proposed by Deng and Lin (2000) and DX generators proposed by Deng and Xu (2003) are as efficient as the classical LCG while achieving an extremely long full period length of $p^k - 1$ and equidistribution property upto k-dimensional space. However, there are two potential problems: (a) their distribution on dimensions higher than k are not the best as indicated by the spectral test discussed in Part II of this book (b) there is a "slow escape from near zero state". This second problem is common with very few non-zero terms used in FMRG and DX generators. For example, with a state k-vector for the generator is "near zero vector", the next variate generated is likely to be zero which will cause a long series of zero variates produced. In contrast, DL/DS/DT generators will produce a non-zero variate quickly from "near zero vector". On the other hand, their $(k+1)$ order equations have a similar structure to DX generators with only a few non-zero terms in their higher-order equations which may cause a similar problem. In comparison, DL generators have fewer nonzero terms than DS/DT in their higher-order equations which may lead to "bad" spectral test values.

3.3 General Class of Efficient MRGs Using MCGs

One way to increase the number of non-zero terms in a large order MRG is to consider a special form of fast MCG which can be viewed as k-dimensional "parallel" implementation of an MRG.

3.3.1 Efficient MRG via MCG

This section presents a brief introduction to implementing matrix congruential generators (MCGs) and their profound connection with maximum-period multiple recursive generators (MRGs). More details will be discussed in Sect. 10.4. The exploration not only sheds light on the theoretical foundations but also provides practical insights for efficient computation.

The MCGs, as presented in Sect. 2.2.2, are a generalization of LCGs but with a matrix multiplier:

$$\mathbf{X}_i = \mathbf{B}\mathbf{X}_{i-1} \bmod p, \ i \geq 1, \tag{3.16}$$

The MCGs' characteristic polynomial determines its period length, with a crucial link to primitive polynomials.

Using the Cayley-Hamilton Theorem, Grothe (1987) gave an MCG implementation of the MRG by choosing multiplier matrix $\mathbf{B} = \mathbf{T}\mathbf{M}_f\mathbf{T}^{-1}$, where $\mathbf{T}$ is an invertible $k \times k$ matrix over $\mathbb{Z}_p^{k \times k}$, and $\mathbf{M}_f$ is the companion matrix to a generator in MRG(k, p):

$$\mathbf{M}_f = \begin{pmatrix} 0 & 1 & 0 & \dots & 0 \\ 0 & 0 & 1 & \dots & 0 \\ \vdots & \vdots & \vdots & \ddots & \vdots \\ 0 & 0 & 0 & \dots & 1 \\ \alpha_k & \alpha_{k-1} & \alpha_{k-2} & \dots & \alpha_1 \end{pmatrix},$$

and it is straightforward to see that

$$f(x) = \det(x\mathbf{I} - \mathbf{M}_f) \bmod p. \tag{3.17}$$

For simplicity, we let MRG(k, p) denote the class of maximum-period k-th order MRGs with prime modulus p. It is shown that, when taken as a k-dimensional vector sequence $(\mathbf{X}_i)_{i \geq 0}$, this MCG satisfies the same recursion as that of the generator in MRG(k, p) with companion matrix $\mathbf{M}_f$; that is, the vector sequence $(\mathbf{X}_i)_{i \geq 0}$ satisfies the following recursion:

$$\mathbf{X}_i = \alpha_1\mathbf{X}_{i-1} + \alpha_2\mathbf{X}_{i-2} + \dots + \alpha_k\mathbf{X}_{i-k} \bmod p, \ i \geq k. \tag{3.18}$$

Therefore, the k sequences taken from each of the k rows in Eq. 3.18 can be viewed as k copies of the same MRG with different starting seeds. Consequently, by employing matrix multiplication, the random number generation based on the MRG can be efficiently realized through the MCG.

3.3.2 Fast MCG (FMCG)

Similar to the method that improves the efficiency of MRG, the efficiency of MCG can be improved by setting as many terms of b_{ij} in MCG to be 0 or ±1 as possible. In particular, Deng and Lin (2000) proposed a special MCG with the maximum period $p^k - 1$ of the following form:

$$\mathbf{B} = \begin{pmatrix} B_1 & -1 & 0 & \cdots & 0 \\ 0 & B_2 & -1 & \cdots & 0 \\ 0 & 0 & . & \cdots & 0 \\ . & . & . & \cdots & . \\ . & . & . & \cdots & . \\ 0 & 0 & 0 & \cdots & -1 \\ -1 & 0 & 0 & \cdots & B_k \end{pmatrix}, \tag{3.19}$$

where $b_i (1 \leq i \leq k)$ are suitably chosen integers. In this case, the matrix in the MCG recursive formula becomes

$$\begin{pmatrix} \mathbf{x}_{new,1} \\ \mathbf{x}_{new,2} \\ \cdots \\ \cdots \\ \mathbf{x}_{new,k} \end{pmatrix} = \begin{pmatrix} B_1\mathbf{x}_1 - \mathbf{x}_2 \\ B_2\mathbf{x}_2 - \mathbf{x}_3 \\ \cdots \\ \cdots \\ B_k\mathbf{x}_k - \mathbf{x}_1 \end{pmatrix} \tag{3.20}$$

Following the naming convention of FMRG, we will call the special form of MCG in Eq. 3.20 or Eq. 3.19 (FMCG).

Clearly, each component of the FMCG defined above is very similar to the LCG defined in Eq. 2.3. The difference between the LCG and each component of the FMCG is that we add a constant increment A in LCG whereas we subtract a variable increment in FMCG. Obviously, we don't expect any difference between the LCG in Eq. 2.3 and the special form of the FMCG in Eq. 3.20 in their computing time. In fact, one can argue that FMCG may be slightly faster than LCG: in order to compute k random numbers, one needs a simple loop computing FMCG whereas one needs to call the LCG routine k times. The overheads of calling the function/subroutine may cause the LCG to be slower than the special form of the FMCG in Eq. 3.20. Like the FMRG, the FMCG has a much longer period than that of LCG. Similar to the discussion of portable MRG, we will restrict $b_i \leq \sqrt{p}$ for the FMCG.

For a general MCG, one can choose the matrix $\mathbf{B}$ so that it can achieve the maximum period of $p^k - 1$. When the order k is large, the number of possible selections of the matrix $\mathbf{B}$ is huge and it could increase the complexity of the generating equations for each component. However, its generating efficiency could be a major concern. Listed below are several issues that need to be addressed:

1. When k is large, it is time-consuming to compute its characteristic function $f_\mathbf{B}(x)$ in Eq. 2.10. One can avoid the actual computation of $f_\mathbf{B}(x)$ by considering a special form of the matrix $\mathbf{B}$ so that $f_\mathbf{B}(x)$ has an explicit formula. In addition, it can be hard to verify that it is a primitive polynomial over $\mathbb{Z}_p$.
2. For a general $k \times k$ matrix $\mathbf{B}$, one may need $O(k^2)$ operations of multiplication and addition and at least $O(k)$ operations of modulus p to produce k variates. Therefore, it can be very time-consuming when k is very large. One can increase the computing efficiency by restricting as many of the entries of $\mathbf{B}$ as possible so that the MCG has the maximum period.
3. A large number of parameters can be helpful to find $f_\mathbf{B}(x)$ to be a primitive polynomial, and it may have a better security property because of increased complexity. However, it could be difficult to list its parameters, and it is usually harder to implement.
4. The proper design choice of MCGs may also depend on the hardware and/or software implementation.

3.3.3 MRGs via FMCG Generators

We can consider a more general form of FMCG with the corresponding matrix whose characteristic polynomial can be computed:

Lemma 3.1 *For a given matrix $\mathbf{B}$ of the form*

$$\mathbf{B} = \begin{pmatrix} B_1 & A_1 & 0 & \cdots & 0 & 0 \\ 0 & B_2 & A_2 & \cdots & 0 & 0 \\ 0 & 0 & B_3 & \cdots & 0 & 0 \\ \vdots & \vdots & \vdots & \ddots & \vdots & \vdots \\ 0 & 0 & 0 & \cdots & B_{k-1} & A_{k-1} \\ A_k & 0 & 0 & \cdots & 0 & B_k \end{pmatrix} \tag{3.21}$$

where A_i and B_i (for $i = 1, \ldots, k$) are integers, the corresponding characteristic polynomial is

$$f_\mathbf{M}(x) = \prod_{i=1}^{k}(x - B_i) - \prod_{i=1}^{k} A_i.$$

For a prime p, if $f_\mathbf{M}(x) = \prod_{i=1}^{k}(x - B_i) - \prod_{i=1}^{k} A_i$ is a k-th degree primitive polynomial over $\mathbb{Z}_p$, then any permutation of A_i and any permutation of B_i (for $i = 1, \ldots, k$) will remain to be the same. If all coefficients are different, then there are upto $k!^2$ possible such combinations. The corresponding matrix $\mathbf{B}$ will yield a maximum period MCG with the same k-th degree primitive polynomial over $\mathbb{Z}_p$. We can explore this feature to increase the "complexity" of the generating equations against the potential attacks.

To minimize the number of parameters required and increase the generating efficiency, we can consider a special form of FMCG as shown below:

Theorem 3.1 *For a given matrix* **B** *of the form*

$$
\mathbf{B} = \begin{pmatrix}
2^{c_1} & 1 & 0 & \cdots & 0 & 0 \\
0 & 2^{c_2} & 1 & \cdots & 0 & 0 \\
0 & 0 & 2^{c_3} & \cdots & 0 & 0 \\
\vdots & \vdots & \vdots & \ddots & \vdots & \vdots \\
0 & 0 & 0 & \cdots & 2^{c_{k-1}} & 1 \\
1 & 0 & 0 & \cdots & 0 & 2^{c_k}
\end{pmatrix}
\tag{3.22}
$$

where b_i (for $i = 1, \ldots, k$) are integers, the corresponding characteristic polynomial is

$$
f_{\mathbf{M}}(x) = \prod_{i=1}^{k}(x - 2^{c_i}) - 1.
$$

$$
\begin{pmatrix}
X_{i,1} \\
X_{i,2} \\
X_{i,3} \\
\vdots \\
X_{i,k}
\end{pmatrix}
=
\begin{pmatrix}
2^{c_1} X_{i-1,1} + X_{i-1,2} \\
2^{c_2} X_{i-1,2} + X_{i-1,3} \\
2^{c_3} X_{i-1,3} + X_{i-1,4} \\
\vdots \\
2^{c_k} X_{i-1,k} + X_{i-1,1}
\end{pmatrix}
=
\begin{pmatrix}
(X_{i-1,1} << c_1) + X_{i-1,2} \\
(X_{i-1,2} << c_2) + X_{i-1,3} \\
(X_{i-1,3} << c_3) + X_{i-1,4} \\
\vdots \\
(X_{i-1,k} << c_k) + X_{i-1,1}
\end{pmatrix}
\quad \mathrm{mod}\ p, \quad i \geq 1.
$$

$$
\tag{3.23}
$$

3.3.4 MRGs via DW Generators

For the sake of computational efficiency, it's common to explore specific MRGs characterized by simple structures with a minimal number of non-zero terms, thereby necessitating fewer resource-intensive multiplications. However, such MRGs often lack desirable properties under spectral testing when compared to their more generalized counterparts, which boast a greater number of non-zero terms.

Conversely, employing generalized MRGs with numerous non-zero terms presents two potential challenges: firstly, ensuring an efficient implementation, and secondly, devising an effective parallelization scheme. Implementing generalized MRGs of higher orders k, can prove challenging due to the substantial number of costly multiplications required by the k-th order linear recurrence to generate subsequent numbers. Additionally, for parallelization schemes, traditional methods like the "jump-ahead parallelization method" for generalized MRGs become computationally inefficient, especially for large values of k.

The DW generator, as proposed by Deng et al. (2023), addresses these challenges by efficiently implementing maximum period MRGs with numerous non-zero terms in parallel, achieved by utilizing an MCG constructed from the MRG. Notably, this approach facilitates the efficient and parallel implementation of a specialized class of high-order MRGs with numerous non-zero terms. Further elaboration on this topic will be provided in Chap. 9.

Computer Search of Maximal-Period Large-Order Multiple Recursive Generators

4

This chapter provides a detailed exploration of computational methods for discovering maximal-period large-order multiple recursive generators (MRGs). We begin with an introduction to Algorithm AK in Sect. 4.1, followed by an examination of the process for searching maximal-period large-order MRGs in Sect. 4.2. Subsequently, attention is directed towards Algorithm GMP in Sect. 4.2.1, which involves investigating generalized Mersenne primes (GMP) and outlining the steps for identifying them. The discussion further extends into the theoretical underpinnings and background of GMP in Sect. 4.2.3, covering topics such as checking primitive polynomials and the theoretical aspects related to prime factors. Together, this chapter provides a comprehensive exploration of computational techniques and theoretical frameworks essential for understanding and finding maximal-period large-order MRGs.

4.1   Algorithm AK: Classical Search Algorithm

A multiple recursive generator (MRG) computes the next value iteratively based on the k most recent values. An MRG with the maximum period is known to have a period length $(p^k - 1)$ with an equidistribution property over a higher dimension. We use $f(x)$ to denote a k-th degree polynomial in Eq. 4.1 whose coefficients are integers between 0 and $p - 1$, inclusively.

$$f(x) = x^k - \alpha_1 x^{k-1} - \cdots - \alpha_k \tag{4.1}$$

We are only interested in polynomials like $f(x)$ which is a monic polynomial where the coefficient of its leading term is one. Furthermore, all the constants and all arithmetic

© The Author(s), under exclusive license to Springer Nature Switzerland AG 2025   43
L. Deng et al., *Random Number Generators for Computer Simulation and Cyber Security*, Synthesis Lectures on Mathematics & Statistics,
https://doi.org/10.1007/978-3-031-76722-7_4

operations (addition, subtraction, negation, multiplication, division, and inversion) are performed modulo a prime number p. Finally, factorization of a kth degree polynomial and its irreducibility check are also performed modulo p (See, Knuth 1998, Niederreiter 1986, for more details). The following are some properties related to modular operation.

Theorem 4.1 (Necessary and sufficient condition of primitive element) *an integer A is a primitive element mod p if the smallest positive exponent e such that $A^e = 1 \bmod p$ is $e = p - 1$.*

A necessary and sufficient condition for an integer A to be a primitive element mod p is that

$$A^{(p-1)/q} \neq 1 \bmod p$$

for any prime factor q of p $- 1$.

Theorem 4.2 (Euler totient) *The number of primitive elements mod p between 0 and $p - 1$ is $\phi(p - 1)$, where $\phi(x)$ is the Euler "totient" function of x, which is the number of integers between 1 and x that are relatively prime to x. It is well-known that if $x = p_1^{r_1} p_2^{r_2} \cdots p_c^{r_c}$, then*

$$\phi(x) = x \prod_{i=1}^{c} \left(1 - \frac{1}{p_i}\right).$$

There are exactly $\phi(p^k - 1)/k$ primitive polynomials of degree k (see, e.g., Knuth 1998).

4.1.1   Three Conditions of Algorithm AK

To search for a maximum period MRG, we need to check whether or not $f(x)$ in Eq. 4.1 is a primitive polynomial. Currently, the common search algorithm is based on the paper by Alanen and Knuth (1964) as described in Knuth (1998, p. 30):

Algorithm 1: Algorithm AK

---

1. Check the following three conditions sequentially. If the given condition is met, then go to the next step; otherwise stop.

    (i). $(-1)^{k-1}\alpha_k$ must be a primitive element mod p.
    (ii). $x^R = (-1)^{k-1}\alpha_k \bmod f(x)$, where $R = (p^k - 1)/(p - 1)$.
    (iii). For each prime factor q of R, the degree of $x^{R/q} \bmod f(x)$ is positive.

2. If all the above conditions are met, then $f(x)$ is a primitive polynomial.

4.1.2   Two Bottlenecks of Algorithm AK

Checking Step (i) in Algorithm AK is fast and its computation complexity does not depend on the size of k. It can reduce the percentage of searches to $\phi(p-1)/(p-1)$.

However, checking Step (ii) is time-consuming even with the most efficient ways of evaluating $x^R \bmod f(x)$. The complexity for exponentiation is of order $O(k)$ and the complexity of k-th degree polynomial multiplication operations is of order $O(k^2)$. The percentage of $f(x)$ satisfying the condition in Step (ii) is less than the percentage of primitive polynomials which is roughly about $1/k$. Furthermore, the number of computations required for a successful or unsuccessful search is of the same order.

Checking Step (iii) is also time-consuming. However, Step (ii) will filter out most of the non-primitive polynomials before reaching Step (iii). The complexity of each iteration in Step (iii) is about the same as in Step (ii) and the number of iterations is C, the number of prime factors in $R = (p^k - 1)/(p - 1)$. The size of C has no apparent relation with the size of R or the size of k. When $C = 1$, then R is a prime number. In that case, Step (iii) is trivial.

When R is not a prime number, the complete factorization of $p^k - 1$ becomes another major bottleneck in Algorithm AK. In practice, it is most common to choose $p = 2^{31} - 1$ which is the largest prime representable as an integer in a 32-bit computer word. Another advantage of choosing such p is that a very efficient operation, without using the actual division, for modulo p is available only for p of the form $2^e - 1$. For this p and $k = 102$ and 120, Deng and Xu (2003) report a successful factorization of $p^k - 1$. Consequently, they find various generators in DX-120-s ($s \le 4$) with period length of $p^{120} - 1 \approx 0.678 \cdot 10^{1120}$. In general, it appears that it is very hard to extend their approach to a much larger value of k.

As previously mentioned, a major bottleneck in our search for higher order MRG is the complete factorization of $p^k - 1$. While choosing $p = 2^{31} - 1$ has some advantages, there is no particularly strong reason that we cannot choose another prime number p. Indeed, removing the restriction on $p = 2^{31} - 1$ can enable us to find a much larger k for which $(p^k - 1)/(p - 1)$ is a prime number. Clearly, k has to be a prime number. Next, we will formally define this class of prime numbers as Generalized Mersenne Prime (GMP).

4.2   Algorithm GMP: Improved Search Algorithm

Large-order MRGs with large periods have gained popularity for their capacity to swiftly generate high-quality random variates, owing to their extensive period lengths, equidistribution in high dimensions, and impressive empirical performance. It is well-known that the maximum period of an MRG is $p^k - 1$, which is reached if its characteristic polynomial

$$f(x) = x^k - \alpha_1 x^{k-1} - \cdots - \alpha_k, \tag{4.2}$$

is a primitive polynomial. When $k = 1$, MRG is a linear congruential generator (LCG) as proposed by Lehmer (1951). In this case, $f(x) = x - B$ is a primitive polynomial of degree one whenever B is a primitive root in a finite field of order p. Consequently, it is desirable to determine coefficients for MRGs of substantial orders that guarantee the attainment of the maximum period. A set of necessary and sufficient conditions for $f(x)$ to be a primitive polynomial has been provided by Knuth (1998, p. 30). One nice property for a maximum period MRG is that it is equidistributed up to k dimensions as in Lidl and Niederreiter (1994, Theorem 7.43). So, every m-tuple ($1 \leq m \leq k$) of integers between 0 and $p - 1$ appears exactly the same number of times (p^{k-m}) over its entire period $p^k - 1$, with the exception of the all-zero tuple which appears one time less ($p^{k-m} - 1$).

Definition 4.1 (*Mersenne Prime (MP)*) The prime numbers of the form $2^k - 1$, where $k \in \mathbb{N}$, are called Mersenne primes

The question of whether there exist infinitely many k values for which $2^k - 1$ is a prime remains an open problem. Presently, only 51 Mersenne Primes are known, with just 17 discovered since 1996. Additional information and ongoing research can be found at the Great Internet Mersenne Prime Search website: https://www.mersenne.org/.

Example 4.1 (*Commonly used MP*)

There are several Mersenne prime numbers of interest:

1. the 8th Mersenne prime number is $2^{31} - 1$, which is the most common prime modulus p for 32-bit computer system;
2. the 24th Mersenne prime number is $2^{19937} - 1$, which is the prime number on which MT19937 is based;
3. the 51th Mersenne prime number is $2^{82589933} - 1$, which is currently the largest one found so far.

A significant bottleneck impeding the efficiency of the AK algorithm in discovering large-order MRGs is the factorization of a substantial number such as $p^k - 1$. L'Ecuyer et al. (1993) introduced a method to overcome this challenge for $k \leq 7$, and later, L'Ecuyer (1999) extended the approach for $k \leq 13$. The method hinges on identifying a prime number p such that $R(k, p) = (p^k - 1)/(p - 1)$ is itself a prime number.

Definition 4.2 (*Generalized Mersenne Prime (GMP)*) For two prime numbers p and k, if

$$R(k, p) = (p^k - 1)/(p - 1)$$

is also a prime number, then $R(k, p)$ is called a Generalized Mersenne Prime (GMP).

Example 4.2 (*GMP when p = 2*)

When $p = 2$, a GMP $R(k, 2)$ is indeed a Mersenne prime number. There are only 51 known values of k for which $R(k, 2)$ is a prime number. They are listed in Table 4.1 below.

While a GMP generalizes the concept of an MP, the objective of GMPs differs. For a known prime k, the aim is to find a prime p such that $R(k, p) = (p^k - 1)/(p - 1)$ is a prime. For a 32-bit RRNG, given a prime k, the task is to find a constant c such that $p = 2^{31} - c$, and the resulting $R(k, p)$ are primes. An additional condition is occasionally imposed on p, requiring that $Q = (p - 1)/2$ is also a prime. While it is difficult to find MPs, it is relatively easy to find a GMP for any prime k.

For 32-bit computers, it is common to choose prime modulus $p = 2^{31} - 1$ because it is the largest (signed) integer that can be stored in a 32-bit computer word and its (expensive) modulus operation can be implemented with efficient logical operations as explained in Deng (2005). In the same paper, he studied an interesting question: What if we fix modulus $p = 2^{31} - 1$ and search for prime order k such that $(p^k - 1)/(p - 1)$ is a prime? According to Deng (2005), there is *no* $k \leq 60,000$ for which $R(k, p)$ is a prime. However, he did find several k ($k = 47$, $k = 643$, $k = 1597$) such that $R(k, p) = q \times H$, where q is a product of several "small" primes (say, $< 10^9$) and H is a (huge) prime. Therefore, such $p^k - 1$ can be completely factorized. To characterize the possible factor of $R(k, p)$, Deng et al. (2012b) applied Theorem 4.3 below which is a special case of Legendre's Theorem for the prime factors of the form $a^k \pm b^k$:

Theorem 4.3 *Let k be a prime. For any prime factor q of $(p^k - 1)/(p - 1)$, $q = 1 \mod 2k$.*

Table 4.1 List of k for all 51 known Mersenne prime numbers, $2^k - 1$

List of k					
2	107	9689	216091	24036583	82589933
3	127	9941	756839	25964951	
5	521	11213	859433	30402457	
7	607	19937	1257787	32582657	
13	1279	21701	1398269	37156667	
17	2203	23209	2976221	42643801	
19	2281	44497	3021377	43112609	
31	3217	86243	6972593	57885161	
61	4253	110503	13466917	74207281	
89	4423	132049	20996011	77232917	

Therefore, any prime factor, say q, of $(p^k - 1)/(p - 1)$ is of the form $q = 2ck + 1$, for some integer c. When k is large, this result is useful to greatly speed up the search of possible factors by skipping all prime numbers $q \neq 1 \bmod 2k$ as possible factors.

Example 4.3 (*k = 7499 and k = 20897*)

With the help of the result $q = 2ck + 1$ and an extensive computer search, Deng et al. (2012b) found the complete factorization of $p^k - 1$ for $k = 7499$ and $k = 20897$.

With the complete factorization found, several efficient MRGs with modulus $(p = 2^{31} - 1)$ are found for orders 102 and 120 in Deng and Xu (2003). Several efficient and portable generators of order $k = 1597$ with period length of approximately $10^{14903.1}$ were constructed in Deng (2005). More recently, Deng et al. (2012b) further extended this work by finding DX and several other classes of generators of order $k = 7499$ and $k = 20897$. It remains an open question whether, with $p = 2^{31} - 1$, there is *any* k such that $(p^k - 1)/(p - 1)$ is a GMP.

4.2.1   Early Exit Strategy and Primality Checking

Deng (2004) proposed an efficient search algorithm for finding maximal period MRGs of large order, called **Algorithm GMP**, that provides an early exit strategy to overcome a second major bottleneck in the algorithm given by Knuth (1998): obtain the complete factorization of $p^k - 1$ when R is not a prime number.

For convenience, we will consider only that $k > 2$. Consequently, k is an odd prime number. For $p = 2$, it is well-known that finding the next $k > 13466917$ is extremely hard. It progresses from 13466917 to 82589933 for 17 years. Instead of searching for k such that $R(k, 2)$ is a prime number, we can search for p for a given k, such that $R(k, p)$ is a prime number. It seems that, for a prime number k, we can find several values of p such that $R(k, p)$ is a GMP. Since our main interest is in 32-bit random number generators, we limit the search to $p < 2^{31}$. The approach has been extended to search for 64-bit or 128-bit random number generators as well. In this paper, for a prime number k, we find the smallest $w > 0$ such that $p = 2^{31} - w$ is a prime number and $R(k, p)$ is a GMP. For any prime number k less than 1000, we have found at least a GMP $R(k, p)$ with $p = 2^{31} - w$. For a larger value of k, the GMP search becomes more and more time-consuming and we limit our search for a k in each interval of one hundred. Again, our search of GMP $R(k, p)$ is successful for k up to 3407. Currently, there is no known theoretical result that we can always find a GMP $R(k, p)$ for any prime number k. As indicated earlier, the main motivation behind the consideration of GMP is to avoid the problem of factoring $(p^k - 1)$. Since our main interest is for large k, say $k > 100$, the computing time for a successful search becomes more important because using Algorithm AK becomes significantly less efficient as k increases. Next, we will turn our attention to the improvement of Algorithm AK.

We summarize the steps of the procedure for finding the proper (p, k) pairs such that $(p^k - 1)/(p - 1)$ is a prime as follows:

Algorithm 2: Obtain the proper (p, k) pairs for GMP

---

1. For a given prime order k, compute Q_n, the product of all prime numbers $q \le n$ of the form $2kc + 1$ with, say, $n = 10^{12}$.
2. For a given prime p, check whether there is any common factor between Q_n and $(p^k - 1)/(p - 1)$:

    a. if the common factor is larger than one (i.e., the early exit condition is satisfied), then we move on to another p;
    b. otherwise, apply the primality tests (see, probabilistic prime test in spscitealtCrandallspsPomerancesps2000) to test the primality of $(p^k - 1)/(p - 1)$:

        i. if the primality test fails, move on to another p;
        ii. otherwise, record the prime modulus p as the result for the current k, and then go on to repeat the whole procedure with the next prime order k.

For the actual search, given a range $[a, b]$ for p and an order k, we followed the steps outlined above to find a prime p where $(p - 1)/2$ is also a prime, a strategy adopted by L'Ecuyer et al. (1993). For 32-bit PRNGs, we started from the upper bound $b = 2^{31} - 1$ and moved downward until we found a suitable prime or reached the lower limit a. Deng (2008) listed some values of p for which $R(k, p)$ is a GMP for k ranging from 5003 to 10007 in increments of around 1000. In this study, using the skipping strategy described earlier, we extended the search and found GMPs for $k = 11003, 12007, 13001, 14009, 15013, 20011$, and 25013. Table 4.2 lists these known GMPs. The number of searches required for each p appeared fairly random and increased with larger values of k.

It is important to note that a GMP differs from a MP, a prime of the form $2^k - 1$. The popular MT19937 generator proposed by Matsumoto and Nishimura (1998), with a period length of $2^{19937} - 1 \approx 10^{6001}$ and equidistribution up to 623 dimensions, is based on a specific MP with $k = 19937$. Although GMPs are extensions of Mersenne primes, their objectives are distinct.

The primary challenge in testing for primitivity modulo p is the factorization of $R(k, p)$. To avoid this issue, one can search for p with a fixed k such that $R(k, p)$ is a prime number, where k must be an odd prime number.

Table 4.2 List of k, c, and $p = 2^{31} - c$, for which $(p^k - 1)/(p - 1)$ is a prime.

k	c	$p = 2^{31} - c$	$\log_{10}(p^k - 1)$
5003	1259289	2146224359	46686.4
6007	9984705	2137498943	56044.7
7001	610089	2146873559	65332.0
8009	5156745	2142326903	74731.1
9001	7236249	2140247399	83983.5
10007	431745	2147051903	93383.7
11003	1276425	2146207223	102676.4
12007	37532781	2109950867	111956.5
13001	292495	2147191153	121323.7
14009	626301	2146857347	130729.2
15013	8996265	2138487383	140072.9
20011	26131941	2121351707	186634.9
25013	11538909	2135944739	233361.1

With the newly found (p, k) pairs of prime modulus p and the associated order k through finding GMPs, we employ Algorithm GMP to search for efficient and portable maximum-period MRGs within the DX/DL/DS/DT classes, as discussed in Chap. 3. The Algorithm GMP is as follows,

---

Algorithm 3: Obtain the proper (p, k) pairs for GMP

---

1. For a given prime order k, compute Q_n, the product of all prime numbers $q \leq n$ of the form $2kc + 1$ with, say, $n = 10^{12}$.
2. For a given prime p, check whether there is any common factor between Q_n and $(p^k - 1)/(p - 1)$:

   a. if the common factor is larger than one (i.e., the early exit condition is satisfied), then we move on to another p;
   b. otherwise, apply the primality tests (see, probabilistic prime test in spscitealtCrandallspsPomerancesps2000) to test the primality of $(p^k - 1)/(p - 1)$:

      i. if the primality test fails, move on to another p;
      ii. otherwise, record the prime modulus p as the result for the current k, and then go on to repeat the whole procedure with the next prime order k.

Even if probabilistic primality tests give a rare false positive for a specific choice of k and p, their impact on finding a primitive polynomial using Algorithm GMP is negligible. Once $R(k, p)$ is identified as a probable prime, it is reasonably safe to assert that the primitive polynomial found subsequently by Algorithm GMP is indeed primitive, ensuring a maximum-period MRG. As Deng (2008) argued, "Once $R(k, p)$ is identified as a probable prime, it is fairly safe to claim that the primitive polynomial found subsequently by Algorithm GMP is indeed primitive, and hence a maximum-period MRG is found."

4.2.2 Empirical Illustration

Algorithm GMP provides an efficient method for finding primitive polynomials due to an early exit strategy since a failed search is likely to terminate very early. To demonstrate this, we show a typical search result for an efficient MRG to be discussed later called a DX-k with $k = 3803$ using Algorithm GMP. In total, the number of iterations needed for the successful search is 2858. For the 2857 failed searches, we tabulate the frequency of early exit for searching DX-3803 at the i-th iteration ($i = 1, 2 \ldots, 1902$) in Table 4.3.

From Table 4.3, one can see that more than 90% of the failed searches will exit at $i \leq 6$ iterations with the average number of iterations equal to 4.997. Because of the early exit, lots of computing time can be saved. Of course, the number of iterations needed until a successful search is "random" but the example given is typical.

Compared with Algorithm AK of Knuth (1998), algorithm GMP has been shown empirically to shorten search time as much as 1000-fold when the order of the MRG is large ($k > 5000$). Using Algorithm GMP, many primitive polynomials of degree up to 25013 corresponding to MRGs of periods up to $\approx 10^{233361}$ have been found. The efficient algorithm GMP can also be applied to find MRGs for 64-bit and 128-bit computers as well. With increases in computing power and drastic reductions in memory costs, the search for ever larger order MRGs with maximal periods and equidistribution over higher dimensions is even more practical.

In summary, Algorithm GMP has the following advantages:

1. It avoids the factorization of $p^k - 1$ by searching for p such that $(p^k - 1)/(p - 1)$ is a prime number. It is well-known that the problem of primality testing is much easier than factorization.
2. It provides an early exit strategy for a failed search, which enables a quick exit in most cases except where a search is successful.

As CPU architectures are moving from 32 bits to 64 bits (or beyond), there is a need for new PRNGs that can increase the resolution of the simulated sequence. Greater precision may be of value in some of the more challenging research problems that arise. To meet this need, the search for large-order MRGs with prime modulus p of magnitude around the

Table 4.3 Frequency (#) and cumulative percentage (c%) of early exit for searching DX-3803 at i-th iteration.

i	#	c%	i	#	c%	i	#	c%	i	#	c%
1	1787	0.6253	20	4	0.9724	39	2	0.9867	111	2	0.9941
2	399	0.7649	21	2	0.9731	40	1	0.9871	112	1	0.9944
3	192	0.8320	22	3	0.9741	41	1	0.9874	124	1	0.9948
4	97	0.8660	23	4	0.9755	44	1	0.9878	128	2	0.9955
5	77	0.8929	24	3	0.9766	45	1	0.9881	134	1	0.9958
6	46	0.9090	25	2	0.9773	48	1	0.9885	145	1	0.9962
7	32	0.9202	26	5	0.9790	50	1	0.9888	163	1	0.9965
8	20	0.9272	27	1	0.9794	54	1	0.9892	170	1	0.9968
9	20	0.9342	28	2	0.9801	55	1	0.9895	173	1	0.9972
10	18	0.9405	29	3	1.9811	56	1	0.9899	176	2	0.9979
11	20	0.9475	30	1	0.9815	58	1	0.9902	217	1	0.9982
12	11	0.9514	31	3	0.9825	59	2	0.9909	255	1	0.9986
13	10	0.9549	32	1	0.9829	62	1	0.9913	296	1	0.9990
14	10	0.9584	33	2	0.9836	63	1	0.9916	563	1	0.9993
15	9	0.9615	34	1	0.9839	68	1	0.9920	587	1	0.9996
16	12	0.9657	35	1	0.9843	74	1	0.9923	1902	1	1.0000
17	5	0.9675	36	1	0.9846	90	1	0.9927			
18	6	0.9696	37	1	0.9850	106	1	0.9930			
19	4	0.9710	38	3	0.9860	108	1	0.9934			

largest integer for a 64-bit or 128-bit word must be continued and expanded. The search algorithms for efficient multipliers in maximal order MRGs may need to be improved.

4.2.3   Theory Behind Algorithm GMP

Our improved search algorithm is based on a combination of some simple but useful results as described next.

Theorem 4.4 *Let* $R(k, p) = (p^k - 1)/(p - 1)$ *be a GMP. Then,* $f(x) = x^k - \alpha_1 x^{k-1} - \cdots - \alpha_k$ *is a primitive polynomial if and only if*

1. *α_k is a primitive element modulo p.*
2. *$f(x)$ is irreducible.*

When $p = 2$, the first condition of Theorem 4.4 becomes redundant. Furthermore, it is known that when $2^k - 1$ is a Mersenne prime number, then a k-th degree irreducible polynomial $f(x)$ is a primitive polynomial. To check quickly whether $f(x)$ is irreducible, we can use Theorem Theorem 4.5 (Crandall and Pomerance 2006, p. 88).

Theorem 4.5 (Irreducibility) $f(x)$ *is irreducible if and only if*

$$\gcd(f(x), x^{p^i} - x) = 1, \text{ for each } i = 1, 2, 3, \ldots, \lfloor k/2 \rfloor.$$

Example 4.4 (*Reducing computations*)

We can use the simple fact that $g(x)^p \bmod f(x) = g(x^p) \bmod f(x)$ and store $(x^p)^j \bmod f(x)$ for $j = 1, 2, \ldots, k$ to reduce the number of computations.

Here, we discuss the major reduction of the average time to find a primitive polynomial using Algorithm GMP. Recall that, for Algorithm AK, the complexity of a successful search and an unsuccessful search are about the same order. Furthermore, the chance of a successful search is only of order $O(1/k)$. Roughly speaking, for Algorithm GMP, the complexity of a successful search is about the same order as Algorithm AK. The main difference is in the time spent on most of the unsuccessful searches. The key point is that the loop $(i = 1, 2, 3 \ldots, \lfloor k/2 \rfloor)$ in Step (ii) of Algorithm GMP will be terminated very early for a large percentage of $f(x)$ tested. Only those k-th degree irreducible polynomials will go through the whole loop. Instead of giving a rigorous argument, it is informative to illustrate this by the following example.

Example 4.5 (*DX-1511-2 PRNGs*)

For $k = 1511$, we first find $p = 2^{31} - 55719 = 2147427929$ such that $R(k, p)$ is indeed a GMP. To search for a generator in DX-1511-2, we use Algorithm GMP to search for B such that

$$f(x) = x^{1511} - Bx^{1510} - B$$

is a primitive polynomial. We start from $B \leq 2^{20}$ downward and we check only B which is a primitive element. After 1866 tries, we find $B = 1044771$ for which $f(x)$ is indeed a primitive polynomial. We can then tabulate the frequency distribution of the loop index i in Step (ii) of Algorithm GMP for this search. From such data, we can see that about 90% of the searches will stop at $i \leq 5$ in Step (ii) while only about 1% of the searches will stop after $i \geq 46$. The average number of iterations is 4.51 with a standard deviation of 29.6. The minimum number of iterations for a (failed) search is 1, while only the successful search requires the maximum 756 iterations.

Primality test for a large integer

Like the irreducible polynomial check, there are two types of integer primality tests. The first type is a probabilistic test which can be highly efficient but there is a tiny probability of making an error. The second type is a deterministic test which can be time-consuming but the conclusion is definite. There is no general polynomial-time algorithm available until Agrawal et al. (2004), known as the AKS algorithm. However, the AKS algorithm is still not yet practical for a large prime number.

Next, we describe our effort to find p for a given prime order k that $R(k, p)$ is a probable prime using the randomized test. For each prime order k, we first find a prime modulus p for probable-prime of $R(k, p) = (p^k - 1)/(p - 1)$. We then verify the primality of $R(k, p)$ using probabilistic tests via some commercial packages such as *Maple* and *MATHEMATICA*. We further perform *industrial prime test* as proposed in Damgard et al. (1993), Crandall and Pomerance (2006). They also discussed that the probability of making a false positive error can be made to be much smaller than 10^{-200}. This error probability is much smaller than the computer software error or the hardware error. Therefore, it can be safely accepted as a "prime" in all but the most sensitive practical applications.

In addition to choosing p for which $(p^k - 1)/(p - 1)$ is also a prime, we require that both p and $Q = (p - 1)/2$ are prime numbers. Here, Q is commonly called a Sophie-Germain prime number.

Effect of Primality test on Search Algorithm

Even if an unlikely mistake were made, for a specific choice of k and p, it has a negligible effect on the successful search for a primitive polynomial. If $R(k, p)$ is not a prime, then we need to check condition (iii) as required in Algorithm AK. Let us assume that $R(k, p) = H \times Q$, where Q is the "smaller" prime factor. From some elementary number theory, one can see that the chance of non-primitive polynomials satisfying the condition (iii) is roughly proportional to $1/Q$. The current computer factorization programs are capable of finding a factor of 50 (or more) digits. Therefore, in the unlikely event $R(k, p)$ is not a prime number, we only have a tiny chance (say, 10^{-50} or less) to find non-primitive polynomials satisfying the condition (iii). For all practical purposes, once the irreducibility test is passed and $R(k, p)$ is shown as a probable prime, we can safely assume that we have found a primitive polynomial.

Practice and Theory of Combination Generators   5

This chapter discusses the importance and benefits of combined generators in pseudo-random number generation, a key aspect of many scientific simulations. The focus is on enhancing the performance of basic generators like Linear Congruential Generators (LCGs) and Multiple Recursive Generators (MRGs), with a particular emphasis on MRGs due to their efficiency and popularity.

LCGs, which were proposed by Lehmer (1951), were once very popular. The drawbacks of LCGs, such as their relatively short cycles, questionable empirical performances, and lack of higher-dimension uniformity, are discussed. These limitations have led to the rise of MRGs, which have become the most popular random number generators in recent years.

The concept of combined generators is then explored. These improve the performance of generators by combining two or more pseudo-random sequences. The first such generator was proposed by Wichmann and Hill (1982), who suggested adding random numbers generated by three simple LCGs. Later, L'Ecuyer (1996) recommended a generator that combines two MRGs of order 3, demonstrating that the combined generator is equivalent to an MRG with a large composite modulus and large coefficients.

This chapter provides theoretical justifications and heuristic arguments for the effectiveness of combined generators. It notes that MRGs can be seen as a special kind of combined generator, and these justifications support the use of MRGs with many nonzero terms. However, it also acknowledges that a general form MRG is not computationally efficient

5.1   Wichmann and Hill Generator

The performance of some generators can be improved by combining two or more pseudo-random sequences. Wichmann and Hill (1982) proposed the first combined generator that produced sequences by adding random numbers generated by three simple LCGs and then taking the fractional part as the random number. Specifically, three simple LCGs considered by Wichmann and Hill (1982) are:

1. $X_i = 171X_{i-1} \bmod 30269 \ (m_1)$,
2. $Y_i = 172Y_{i-1} \bmod 30307 \ (m_2)$,
3. $Z_i = 170Z_{i-1} \bmod 30323 \ (m_3)$.

The combination generator is

$$U_i = X_i/m_1 + Y_i/m_2 + Z_i/m_3 \bmod 1 \tag{5.1}$$

and its period length is increased to $LCM(m_1 - 1, m_2 - 1, m_3 - 1) \approx 6.95 \times 10^{12}$. Zeisel (1986) showed that the combination generator is equivalent to another LCG with a large multiplier ($B = 16555425264690$) and a large modulus ($p = 27817185604309 = m_1 \times m_2 \times m_3$), which will be discussed in detail in Sect. 5.3.

5.2   Lecuyer's MRG32k3

L'Ecuyer (1999) proposed a popular generator, called *MRG32k3a*, by combining the following two MRGs, each of order 3:

$$X_i = 1403580X_{i-2} - 810728X_{i-3} \bmod (2^{32} - 209),$$

$$Y_i = 527612Y_{i-1} - 1370589Y_{i-3} \bmod (2^{32} - 22853),$$

in which
$$Z_i = (X_i - Y_i) \bmod (2^{32} - 209).$$

The corresponding uniform (0,1) generator is

$$U_i = Z_i^*/(m_1 + 1), \quad m_1 = 2^{32} - 209, \tag{5.2}$$

where $Z_i^* = Z_i$, if $Z_i > 0$ and $Z_i^* = m_1$, if $Z_i = 0$. The period length is about 3×10^{57}.

The idea of combination generators is not limited to combining PRNGs of the same type. Marsaglia's Super-Duper generator was one of the earliest generators (see, e.g., Learmonth and Lewis 1973) to combine two different types of PRNGs. The Super-Duper generator combined, by exclusive OR, an LCG, and a linear feedback shift register to obtain a generator with a period length of around 2^{62} that performed better than either of its component generators alone. Super-Duper was used early on in several software packages, including S-PLUS and it is still an option in a random number generator function in R.

While the procedure of combining generators can generally improve the empirical performances of the PRNGs and increase the period length, it will greatly decrease the generating efficiency and it will not yield a PRNG with an exact high-dimensional equi-distributional property.

5.3   Chinese Remainder Theorem

The Chinese Remainder Theorem is fundamental in number theory with applications in various areas of mathematics and computer science, particularly in cryptography. It provides a solution to simultaneous modular congruences. Suppose we have a set of congruences:

$$x \equiv a_1 \pmod{m_1}$$
$$x \equiv a_2 \pmod{m_2}$$
$$\vdots$$
$$x \equiv a_n \pmod{m_n}$$

where $a_1, a_2, \ldots, a_n$ are integers and $m_1, m_2, \ldots, m_n$ are pairwise coprime. Then, there exists a unique solution for x modulo M, where M is the product of all the moduli $m_1, m_2, \ldots, m_n$. This solution can be found using the following formula:

$$x \equiv \sum_{i=1}^{n} a_i M_i y_i \pmod{M}$$

where $M_i = M/m_i$, and y_i is the modular multiplicative inverse of M_i modulo m_i.

For the Wichmann and Hill generator, the equivalent LCG $X_i = B X_{i-1} \bmod p$ can be found using the Chinese Remainder Theorem, which is to solve $B = 171 \bmod m_1$, $B = 172 \bmod m_2$, $B = 170 \bmod m_3$, where $m_1 = 30269$, $m_2 = 30307$, and $m_3 = 30323$. The expression to compute B is obtained as:

$$B = (B_1 + B_2 + B_3) \bmod (m_1 m_2 m_3),$$

where B_1, B_2, B_3 can be found by taking the multiplicative inverses as

$$B_1 = 171(m_2 m_3)((m_2 m_3)^{-1} \bmod m_1),$$

$$B_2 = 172(m_1 m_3)((m_1 m_3)^{-1} \bmod m_2),$$

$$B_3 = 170(m_1 m_2)((m_1 m_2)^{-1} \bmod m_3).$$

5.4   Statistical Justification

Wichmann and Hill (1982) did not provide any theoretical justification for their combination generator. Several theoretical justifications for combined generators have been given in the literature (see, e.g., Brown et al. 1979, Gentle 2003). Specifically, Marsaglia (1985) proved that a combined generator yields a distribution closer to or at least no farther from the uniform distribution than that of the corresponding individual generators.

L'Ecuyer (1997) showed that the set of all t ($> k$) tuples in the generated MRG sequence as given in Eq. 5.2 are distributed in equidistant parallel hyperplanes at distance

$$d_t > \left(1 + \sum_{i=1}^{k} \alpha_i^2\right)^{-1/2}. \tag{5.3}$$

Therefore, a large $\sum_{i=1}^{k} \alpha_i^2$ is a necessary (but not sufficient) condition for a good MRG. Consequently, it is better to have more nonzero terms with larger α_i^2 in general. This gives a nice justification for combining MRGs since a combined MRG is equivalent to another MRG with large nonzero coefficients according to L'Ecuyer (1996). Unfortunately, this justification does not apply to a general combined generator without a simple linear structure. For example, the generating methods for the component generators are not necessarily of the type of MRG or LCG. In the following, we will discuss justifications from the statistical viewpoint. We find that the general conclusions of favoring more nonzero terms from various viewpoints are quite consistent. To support these statistical justifications, the results of an empirical performance study are presented.

Deng and George (1990) provided an additional theoretical justification by showing that a a combined generator should improve the uniformity with the following theorem:

Theorem 5.1 *Let X_i be a continuous random variable on $[0, 1]$ with the probability density function (probability density function) $f_i(x_i)$ such that $|f_i(x_i) - 1| \leq \epsilon_i$ for some constant $\epsilon_i, i = 1, 2, \ldots, n$ and X_i's are independent. Let $Y = \sum_{i=1}^{n} X_i \bmod 1$. Then the probability density function of Y, $f(y)$, satisfies*

$$|f(y) - 1| \leq \prod_{i=1}^{n} \epsilon_i.$$

Therefore, by combining the random variate generators, we can improve the n component (possibly "defective") generators, represented as X_i's ($i = 1, 2, \ldots, n$) whose distributions are close but not exactly uniform. Note the assumption on the existence of the density of X_i ($i = 1, 2, \ldots, n$) is unrealistic. However, we can treat the PRNGs as simply finite-bit realizations of a random variable, X_i with a probability density function.

To generate a variate Y, we normally will first generate X_i's by some random variate generators. It may be unrealistic to assume the variates X_i's generated by these generators are independent of each other. Deng et al. (1997) showed that the independence assumption in Theorem 5.1 can be further relaxed. In particular, they considered an extreme case in which the X_i's are all identical to a random variable X. That is, $Y = nX$ mod 1. They showed that the asymptotic distribution of Y is a $U(0, 1)$ distribution as given in the following theorem:

Theorem 5.2 *Let X be a random variable on $[0, 1]$ with probability density function $f_X(x)$. Let $Y_n = nX$ mod 1. Then the probability density function of Y_n, $f_n(y)$, satisfies*

$$|f_n(y) - 1| \longrightarrow 0, \quad as \ n \longrightarrow \infty.$$

Deng et al. (1997) showed that the independence assumption in Theorem 5.2 can be further relaxed. In particular, they considered an extreme case in which all X_i's are all identical to a random variable X. That is, $Y = nX$ mod 1. They showed that the asymptotic distribution of Y is a $U(0, 1)$ distribution as given in the following theorem:

Theorem 5.3 *Let X be a random variable on $[0, 1]$ with probability density function $f_X(x)$. Let $Y_n = nX$ mod 1. Then the probability density function of Y_n, $f_n(y)$, satisfies*

$$|f_n(y) - 1| \longrightarrow 0, \quad as \ n \longrightarrow \infty.$$

According to the above theorems, one can generate a "flatter" distribution by either adding up several variables or multiplying a random variate with a large constant and then taking its fractional part.

The problem of constructing a set of random digits by "compound randomization" was first studied by Horton (1948), Horton and Smith III (1949a). They showed that adding several pseudo numbers modulo m, where m is a positive integer, will converge to a discrete uniform distribution over $0, 1, 2, \ldots m - 1$. Similar results for continuous random variables can be obtained. It can be shown that the fractional part of the sum of two independent random variables, one of which is U(0,1), will produce a U(0,1) random variable (as in Theorem 5.4).

Theorem 5.4 *Let Y and U be independent random variables such that $U \sim U(0, 1)$. Let $X = Y + U$ mod 1. Then $X \sim U(0, 1)$.*

From Theorem 5.4, we obtain the following:

Corollary 5.1 *Let $Y_1, Y_2, \ldots, Y_n$ be independent random variables distributed on $(0,1)$. Let $X = \sum_{i=1}^{n} Y_i$ mod 1. If $Y_1 \sim U(0, 1)$, then $X \sim U(0, 1)$.*

A modification of the statement of Theorem 5.4 produces the following characterization of the uniform distribution:

Theorem 5.5 *Let U be a random variable with a distribution on (0,1). If X = U + Y mod 1 is uniformly distributed on (0,1) for every continuous random variable Y with support S_Y which is a subset of (0,1), then $U \sim U(0, 1)$, and conversely.*

Thus, one can produce a random variate whose distribution is closer to uniform distribution by taking the fractional part of the sum of several variables. If it is assumed that the generated variates are independent, then these theorems provide a theoretical justification of the algorithm proposed in Wichmann and Hill (1982).

5.5   Intuitive Explanations for Theoretical Justification

We remark here that the previous theories are based on the assumption that the basis variable X is a continuous random variable with a probability density function. In practice, random number generators can only produce a finite number of discrete points in (0, 1), and hence the generated variates are only approximations to realizations of the corresponding continuous random variables. As shown in Theorem 5.4 and later Theorems, the assumption of the existence of a probability density function can yield some exact and simple expression to shed light on the theoretical justification. In fact, these assumptions have been implicitly made on the random variates produced by random number generators when using them. For instance, the inverse CDF method operates under the assumption that the discrete points generated within the interval (0,1) represent random samples derived from a continuous uniform distribution (see, e.g., Example 2.1). In general, whether it's the principles outlined in this theory or other algorithms like the inverse CDF method, the effectiveness relies on component generators capable of producing points that are sufficiently "dense" across the (0, 1) range. Implicit in any generation algorithm for probability distributions is the assumption that we can generate a sequence of independent random variables with a uniform distribution over the range (0, 1). However, in practice, we can only generate a sequence of discrete random variables, the distribution of which ideally closely approximates the uniform (0,1) distribution.

Stretching the probability density function of a continuous random vector beyond the $(0, 1)^k$ cube to a wider domain like $(0, n)^k$ for $n > 1$ results in a "flatter" distribution within each unit subcube. Taking the fractional part of the stretched vector consolidates the pieces back into $(0, 1)^k$, creating a probability density function that is the sum of probability density function's overall subcubes. This yields a more uniformly distributed random vector over $(0, 1)^k$, roughly independent of the original individual random vectors.

Since there are only a finite number of digits representable by a computer, we need to be aware of practices designed to improve the uniformity of an PRNG. The process of "stretching" can be achieved by adding variates or multiplying by a large constant. Multiplying an PRNG by a number maintains its recurrence relationship and lattice structure, but may lose lower-order bits without careful consideration. Combining PRNGs with different modulus increases the period of the resulting PRNG.

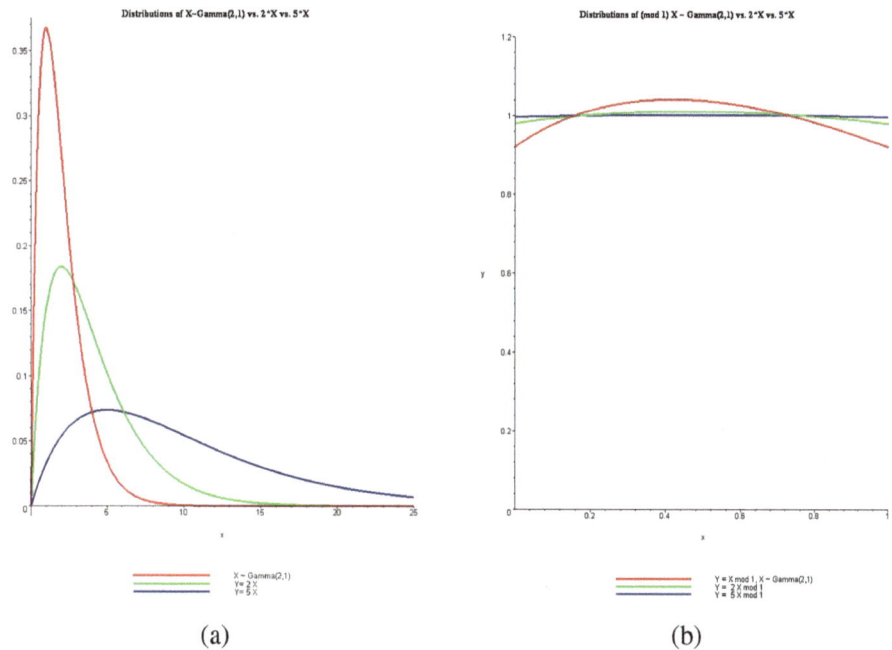

Fig. 5.1 Plots of probability density functions. Left panel is for X (red), $2X$ (green) and $5X$ (blue) with $X \sim Gamma(2, 1)$ and right panel is for $Y = cX$ mod $1, c = 1, 2, 5$

Graphical illustrations in Example 5.1 demonstrate that the probability density function, $f(y)$, for the random variable $Y = \sum_{i=1}^{n} X_i$ mod 1 becomes flatter as the variance of Y increases, regardless of independence among individual components, X_i. Consequently, $f(y)$ flattens, and the "mod 1 operation" accumulates density functions over the unit interval, resulting in a random variable closely approximating $U(0, 1)$.

Example 5.1 (*Intuitive explanation of Theorem 5.1 it in the case of uni-variate*)

As an illustrative example, we consider a random variable X having an exponential distribution with a mean of 1. No one will consider using X mod 1 as a random number generator for $U(0, 1)$.

However, Fig. 5.1 shows a clear trend toward the uniform density as we "stretch out" X. In the left panel of Fig. 5.1, the plots of the probability density functions for X (in red), $2X$ (in green), and $5X$ (in blue) with $X \sim Gamma(2, 1)$, shows that the probability density function for $Y = cX$ becomes flatter as c increases. Next, in the right panel of Fig. 5.1, we plot the probability density functions for $(X$ mod 1) shown in red, $(2X$ mod 1) shown in green, and $(5X$ mod 1) shown in blue with $X \sim Gamma(2, 1)$. It is easy to confirm the above "stretching-and-flatten" argument without formal proof.

The intuitive argument can be extended to k-dimensions. That is, it can be easily extended to the combination of n k-dimensional random vectors with $\mathbf{Y} = \mathbf{X}_1 + \mathbf{X}_2 + \cdots + \mathbf{X}_n$ mod 1 and we can show a similar result for $\mathbf{Y} = n\mathbf{X}$ mod 1, where $\mathbf{X}$ is a k-dimensional random vector. Almost the same argument can be used by stretching (in each of the k dimensions) the joint probability density function which will flatten the "surface" of the joint probability density function. Taking the "mod 1 operation" on the random vector will cause it to converge to a uniform distribution of $U[0, 1]^k$. Theoretical justifications for combining random vectors can be found in Deng (2016), Deng and Chu (1991), Deng et al. (1997).

Example 5.2 (*Intuitive explanation in the case of a bivariate*)

One can also illustrate the above argument using a simple example with (X, Y) mod 1, where (X, Y) has a bivariate normal distribution with a large correlation coefficient $\rho = 0.99$.

Figure 5.2a shows that even after taking "mod 1" operation, the probability density function is not very close to a uniform distribution over $[0, 1]^2$. However, the successive plots of probability density functions for $((2X, 2Y)$ mod 1), $((4X, 4Y)$ mod 1), and $((8X, 8Y)$ mod 1) show the probability density function is indeed approaching a uniform distribution over $[0, 1]^2$.

Although the distributions of both component generators (X_i and C_i) do not closely match the desired uniform distribution, the MWC is expected to exhibit a distribution closer to uniform. While the concept of combining n generators can enhance their performance, it may considerably reduce generating efficiency as n grows. It is imperative to explore alternative classes of efficient generators characterized by extremely long period lengths and favorable distributional properties.

5.6   High Dimensional Property for Combination Generators

Suppose that $\{(U_{j0}, U_{j1}, U_{j2}, \ldots), j = 1, 2, \ldots, n\}$ are n sequences of random variates generated by any PRNG. No assumption is necessary regarding how each PRNG is generated. The combined sequence $\{Y_1, Y_2, \ldots\}$ where $Y_i = U_{1i} + U_{2i} + \cdots + U_{ni}$ mod 1 will be studied from a statistical viewpoint in this section. In particular, for any positive integer k, we would like to investigate the joint probability distribution of the k-dimensional random vector $\mathbf{Y} = \mathbf{X}_1 + \mathbf{X}_2 + \cdots + \mathbf{X}_n$ mod 1, where $\mathbf{X}_j = (U_{ji_1}, U_{ji_2}, \ldots, U_{ji_k})'$ with $i_1 < i_2 < \cdots < i_k$. For a vector $\mathbf{x} = (x_1, x_2, \ldots, x_k)'$, let $\mathbf{x}$ mod 1 $= (x_1 \bmod 1, x_2 \bmod 1, \ldots, x_k \bmod 1)'$.

We will use a k-dimensional random vector $\mathbf{X}$ to represent specific k component realizations of an PRNG. Furthermore, for simplicity, we will assume the existence of the probability density function $f_X(x)$ for X. This nevertheless greatly reduces the need for exact computations with discrete values.

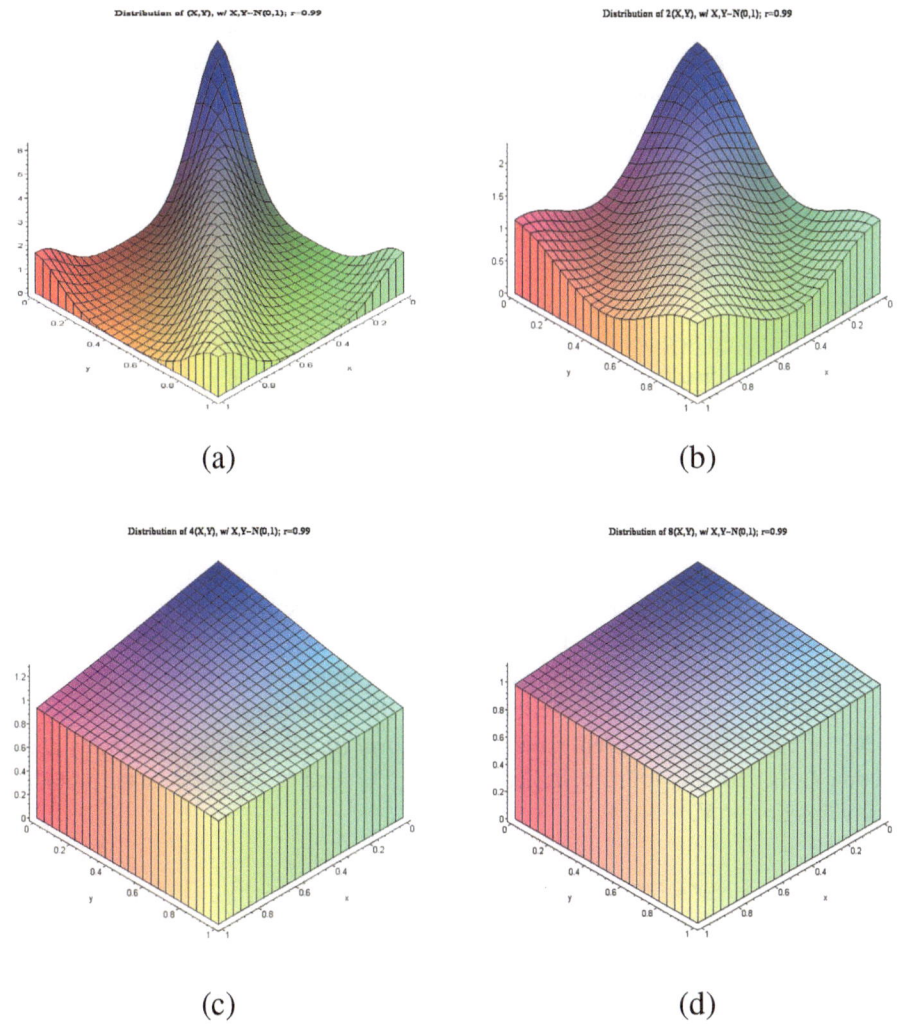

(a)   (b)

(c)   (d)

Fig. 5.2 Plots of probability density functions in the case of a bivariate normal vector (X, Y) with $\rho = 0.99$: **a** (X, Y) mod 1, **b** $(2X, 2Y)$ mod 1, **c** $(4X, 4Y)$ mod 1, and **d** $(8X, 8Y)$ mod 1

For $0 < \epsilon < 1$, let $L_r(\epsilon)$ be the class of probability density function $f_{\mathbf{X}}(\mathbf{x})$ over $[0, 1]^k$ which is in the neighborhood of the uniform probability density function such that

$$||f_{\mathbf{X}}(\mathbf{x}) - 1||_r = \left(\int_{[0,1]^k} |f_{\mathbf{X}}(\mathbf{x}) - 1|^r d\mathbf{x} \right)^{1/r} \leq \epsilon.$$

The value of 1 is so special here because it is the probability density function of $U[0, 1]^k$. Furthermore, we set $0 < \epsilon < 1$, because it is desirable for a vector PRNG to generate a distribution reasonably close to $U[0, 1]^k$.

The following theorem provides a theoretical justification for the goodness of the combination generators. Specifically, the theorem shows that the fractional part of the sum of two independent (nearly) uniform random vectors will produce a distribution whose probability density function is closer to the uniform distribution.

Theorem 5.6 *Let* X_1, X_2 *be any two independent random vectors over* $[0, 1]^k$, *with the probability density functions* $f_{X_1}(x)$ *and* $f_{X_2}(x)$. *Let* $Y = X_1 + X_2$ *mod* 1, *and* $f_Y(y)$ *be the probability density function of* Y. *For any* $r_1, r_2 \geq 1$ *such that* $\frac{1}{r_1} + \frac{1}{r_2} = 1$, *we have*

$$|f_Y(y) - 1| \leq ||f_{X_1}(x) - 1||_{r_1} ||f_{X_2}(x) - 1||_{r_2}.$$

That is, if $f_{X_1}(x) \in L_{r_1}\epsilon_1$ *and* $f_{X_2}(x) \in L_{r_2}\epsilon_2$ *then* $f_Y(y) \in L_\infty\epsilon_1 \cdot \epsilon_2$.

Essentially, we have proved here that the combination generator will not only improve the "uniformity" of the generator but also its "independence" when considering the joint distribution of any k-dimensional random vectors.

According to Theorem 5.6, one can produce a random vector whose distribution is closer to $U[0, 1]^k$ by taking the fractional part of the sum of several random vectors: Let $X_1, X_2, \ldots, X_n$ be n independent random vectors over $[0, 1]^k$, with the probability density function $f_{X_i}(x)$, for $i = 1, 2, \ldots, n$. Let $Y = \sum_{i=1}^n X_i$ mod 1, and $f_Y(y)$ be the probability density function of Y.

- If $|f_{X_i}(x) - 1| \leq \epsilon_i$, for $i = 1, 2, \ldots, n$, then $|f_Y(y) - 1| \leq \prod_{i=1}^n \epsilon_i$.
- If $\prod_{i=1}^n \epsilon_i \longrightarrow 0$, then Y converges to $U[0, 1]^k$
- In particular, if one of the X_i is uniformly distributed over $[0, 1]^k$, then Y is uniformly distributed over $[0, 1]^k$.

In practice, since a sequence generated by any PRNG is deterministic, this independence assumption seems to be unrealistic. To show that the combination generator is indeed superior even when X_i's are dependent, we next consider the extreme case that all X_i are equal to X. In this case, we will show that the distribution of

$$Y = nX \text{ mod } 1$$

will be closer to the uniform distribution than that of each X. In fact, this statement holds for a more general case

$$Y = DX \text{ mod } 1, \quad D - \text{diag}(n_1, n_2, \ldots, n_k), \tag{5.4}$$

where n_j's are positive integers. Furthermore, we will show that for large n_j's, the distribution of Y is nearly distributed as $U[0, 1]^k$.

Let D be defined as in Eq. 5.4 and

$$S_D = \{\mathbf{i} = (i_1, i_2, \ldots, i_k) | 0 \le i_j \le n_j - 1\}.$$

1. Let X be any random vectors over $[0, 1]^k$, with the probability density function $f_X(x)$. Then the probability density function of $Y = DX$ mod 1 is

$$f_Y(y) = \frac{1}{n^k} \sum_{\mathbf{i} \in S_D} F_X(x) D^{-1}(\mathbf{y} + \mathbf{i}).$$

2. For any integrable function $e(\mathbf{x})$ defined over $[0, 1]^k$, we have

$$\int_{\mathbf{x} \in [0,1]^k} e(\mathbf{x}) dx = \frac{1}{n^k} \sum_{\mathbf{i} \in S_D} \int_{\mathbf{y} \in [0,1]^k} e(D^{-1}(\mathbf{y} + \mathbf{i})) dy.$$

Note that $Y = DX$ mod 1 is not a one-to-one transformation of X.

Theorem 5.7 *Let X be any random vectors over $[0, 1]^k$, with the probability density function $f_X(x)$. Let $Y = DX$ mod 1, where D be defined as in Eq. 5.4, and $f_Y(y)$ be the probability density function of Y.*

1. *For any $r \ge 1$, we have*

$$||f_Y(y) - 1||_r \le ||f_X(x) - 1||_r.$$

Therefore, if $f_X(x) \in L_r \epsilon$ then $f_Y(y) \in L_r \epsilon$.

2.

$$|f_Y(y) - 1| \to 0, \quad as \ \min(n_1, n_2, \ldots, n_k) \to \infty.$$

That is, the components of Y will not only "more uniform" but also "more independent" of each other.

3. *Moreover, Y and X will be asymptotically independent of each other.*

Theorem 5.7 presents a comprehensive understanding of stretching continuous random vectors. It indicates a method where stretching X along its axes and extracting the fractional parts leads to a distribution closer to $U(0, 1)$ with enhanced independence among its components. Furthermore, the theorem offers justification for employing a large multiplier in a Multiplicative Linear Congruential Generator (MLCG), demonstrating that the successive variates generated by an MLCG should asymptotically achieve independence and uniform distribution.

Part (1) of Theorem 5.7 shows that "stretching out" any continuous random vector X will be as good as the X itself. Part (2) shows that by stretching X in each directions and taking its fractional part we can obtain a distribution closer to $U(0, 1)$ while each component will be more independent of each other. Part (3) gives some justification for a MLCG with a large multiplier. It shows that the successive variates generated by a MLCG should be asymptotically independently and uniformly distributed.

5.7   Empirical Evaluations for Combination Generators

To study the empirical performance of combined generators, we generate two non-uniform variates X_1 and X_2 independently from the following two beta distributions:

$$X_1 \sim \text{beta}(2,1), \quad X_2 \sim \text{beta}(1,2),$$

by

$$X_1 = U_1^{1/2}, \quad X_2 = 1 - U_2^{1/2},$$

using two good PRNGs, such as the ones discussed later, to generate U_1 and U_2.

For each selection of (n_1, n_2) listed in Table 5.1, we evaluate the empirical performance of

$$Y = n_1 X_1 + n_2 X_2 \quad \text{mod } 1. \tag{5.5}$$

The following three well-known empirical test packages are currently most popular for testing a random number generator (which was detailed discussed in Sect. 2.8:

1. *DIEHARD* test suite: It was developed by Marsaglia [1996] and is available from http://www.stat.fsu.edu/pub/diehard. *DIEHARD* test package is the oldest and most famous test suite. However, it contains relatively few tests.
2. *NIST* test suite: It was developed by National Institute of Standards and Technology and one can download it from http://www.csrc.nist.gov/rng/. Some of the tests are similar to the *DIEHARD* suite.
3. *TestU01* test suite: It was developed by Professor L'Ecuyer and it is by far the most comprehensive test suite. Its source code and user's manual are available from http://www.iro.umontreal.ca/~lecuyer/.

We choose *TestU01* to evaluate the empirical performance of the combined generators considered in our study. There are three predefined test modules in *TestU01*: (i) *Small crush* has the fewest tests (14 or 15, depending on the version used) and takes less than 1 min of computing time. It appears that the conclusions of the empirical evaluation for various versions of TestU01 (v1.0, v1.1 or v1.2) do not change much. (ii) *Crush* has more tests (94 in the version we used) and its running time is less than 2 hrs. (iii) *Big crush* is the most comprehensive module and may require more folds of computing time.

We apply the test module *Crush* to each of the 60 selections of (n_1, n_2) with n_1 ranging from 1 to 71 and n_2 ($\geq n_1$) ranging from 3 to 997. For each selection, *Crush* module produces 94 p-values, one for each of the 94 tests. A significantly small p-value indicates that the generator fails that particular test. We then count the number of tests with p-values less than 10^{-300}, 10^{-20}, and 10^{-4}, respectively.

From Table 5.1, we can see that the count of small p-values tends to decrease as either n_1 or n_2 increases. When both n_1 and n_2 are relatively small, several of the 94 empirical tests reject the generator Y with an extremely small p-value (less than 10^{-300}). As an example, for $(n_1, n_2) = (1, 3)$, 24 tests out of the 94 tests report p-values less than 10^{-300} while $65(= 94 - 29)$ tests report p-values at least 10^{-4}. For $(n_1, n_2) = (7, 31)$, none of the 94 tests report p-values less than 10^{-20} while only 2 tests report p-values less than (or equal to) 10^{-4}.

It is interesting to explore the relationship between test counts of small p-values with a certain function $g(n_1, n_2)$ of n_1 and n_2. Two obvious choices are (1) $L(n_1, n_2) = n_1^2 + n_2^2$ and (2) $L(n_1, n_2) = n_1 \times n_2$. The first choice is motivated by the lattice structure lower bound given in Eq. 5.3 as reported in L'Ecuyer (1997) and the second choice is motivated by Theorems 5.4 and 5.5. We find that the second choice shows a clearer relationship between the two as explained next.

Table 5.1 Small p-value counts for $Y = n_1 X_1 + n_2 X_2 \bmod 1$ versus (n_1, n_2)

n_1	n_2	10^{-300}	10^{-20}	10^{-4}	n_1	n_2	10^{-300}	10^{-20}	10^{-4}	n_1	n_2	10^{-300}	10^{-20}	10^{-4}
1	3	24	27	29	2	101	0	0	2	7	7	3	4	11
1	5	21	24	28	2	307	0	0	0	7	17	2	2	3
1	11	12	15	18	2	997	0	0	0	7	31	0	0	2
1	23	10	11	14	3	5	11	12	18	7	47	0	0	1
1	31	9	10	11	3	7	11	12	14	7	59	0	0	0
1	59	3	4	6	3	23	3	4	8	7	101	0	0	1
1	101	2	2	5	3	43	2	2	4	11	31	0	0	0
1	211	1	1	4	3	59	1	1	3	11	47	0	0	0
1	307	0	0	0	3	97	0	0	2	13	23	0	0	0
1	503	0	0	0	3	101	0	0	2	13	31	0	0	0
1	997	0	0	0	3	211	0	0	0	15	15	0	0	3
2	3	16	20	27	3	991	0	0	0	15	31	0	0	1
2	5	12	16	20	5	7	8	9	12	15	43	0	0	0
2	7	11	13	18	5	19	2	2	4	19	19	0	0	0
2	11	10	11	13	5	31	2	2	2	23	23	0	0	0
2	23	3	4	10	5	43	0	0	2	35	59	0	0	0
2	31	3	3	7	5	59	0	0	1	41	61	0	0	0
2	59	2	2	3	5	97	0	0	0	53	69	0	0	0
2	71	2	2	4	5	101	0	0	0	71	119	0	0	0

Table 5.2 Counts Ordered by $L(n_1 n_2) = \log_{10}(n_1 \times n_2)$

n_1	n_2	10^{-300}	10^{-20}	10^{-4}	n_1	n_2	10^{-300}	10^{-20}	10^{-4}	n_1	n_2	10^{-300}	10^{-20}	10^{-4}
1	3	24	27	29	2	101	0	0	2	7	7	3	4	11
1	5	21	24	28	2	307	0	0	0	7	17	2	2	3
1	11	12	15	18	2	997	0	0	0	7	31	0	0	2
1	23	10	11	14	3	5	11	12	18	7	47	0	0	1
1	31	9	10	11	3	7	11	12	14	7	59	0	0	0
1	59	3	4	6	3	23	3	4	8	7	101	0	0	1
1	101	2	2	5	3	43	2	2	4	11	31	0	0	0
1	211	1	1	4	3	59	1	1	3	11	47	0	0	0
1	307	0	0	0	3	97	0	0	2	13	23	0	0	0
1	503	0	0	0	3	101	0	0	2	13	31	0	0	0
1	997	0	0	0	3	211	0	0	0	15	15	0	0	3
2	3	16	20	27	3	991	0	0	0	15	31	0	0	1
2	5	12	16	20	5	7	8	9	12	15	43	0	0	0
2	7	11	13	18	5	19	2	2	4	19	19	0	0	0
2	11	10	11	13	5	31	2	2	2	23	23	0	0	0
2	23	3	4	10	5	43	0	0	2	35	59	0	0	0
2	31	3	3	7	5	59	0	0	1	41	61	0	0	0
2	59	2	2	3	5	97	0	0	0	53	69	0	0	0
2	71	2	2	4	5	101	0	0	0	71	119	0	0	0

Table 5.2 sorts the test counts of small p-values by $n_1 \times n_2$, or equivalently, by $\log_{10}(n_1 \times n_2)$. From Table 5.2, we can clearly see that the count of small p-values tends to decrease as $n_1 \times n_2$ increases. When $\log_{10}(n_1 \times n_2) \geq 2.33$ (or $n_1 \times n_2 \geq 214$), none of the 94 tests report p-values of 10^{-20} or less while two tests report p-values of 10^{-4} or less. When $\log_{10}(n_1 \times n_2) \geq 3$ (or $n_1 \times n_2 \geq 1000$), no test reports p-values of 10^{-4} or less. The relationship between the counts of small p-values versus $\log_{10}(n_1 \times n_2)$ as listed in Table 5.2 can be clearly seen in Fig. 5.3.

The main purpose of this empirical study is to show the applicability of the theory developed from a statistical justification viewpoint. We have successfully demonstrated that one can easily transform any "bad" uniform random number generators into "good" uniform random number generators by linearly combining these generators. The same theoretical justification leads us to the consideration of MRGs with many nonzero terms while maintaining the efficiency and portability of the MRGs.

Remark 5.1 Employing statistical theory, we have provided rationales for the utilization of combined generators. Notably, we have demonstrated the applicability of these justifications to recently introduced generators, specifically the DL and DS generators. Beyond the statistical underpinnings outlined here, these generators boast exceptionally long period

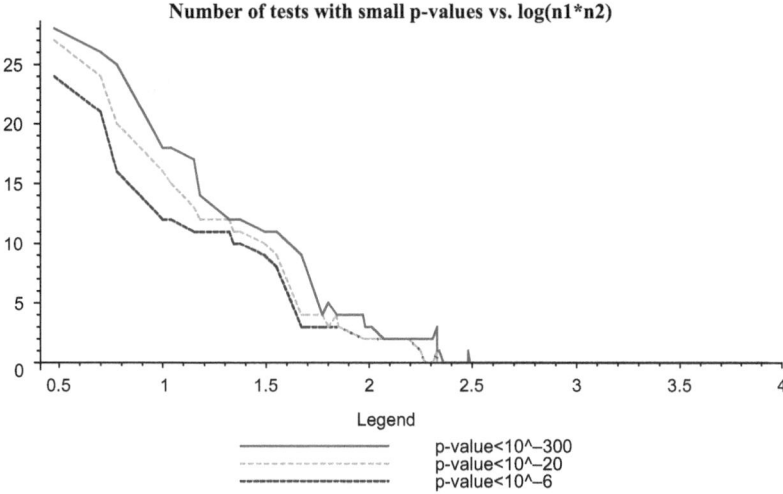

Fig. 5.3 Plot of test counts with small p-values versus $\log_{10}(n_1 \times n_2)$

lengths and possess a high-dimensional equi-distribution property. The DX, DL, and DS generators emerge as robust alternatives to MT19937, a well-known generator proposed by Matsumoto and Nishimura (1998). MT19937 operates on a linear recurrence of order $k = 19937$ modulo $p = 2$, featuring a period of $2^{19937} - 1 \approx 10^{6001.6}$ and equidistribution up to 623 dimensions.

5.8   Period Length for Combination Generators

Consider two random-variate sequences, X_i's and Y_i's, generated by two generators X and Y over $\mathbb{Z}_m$, respectively. Let $Z = (X + Y) \bmod m$, a combination generator of X and Y, and Z_i's be the associated sequence. Let P_x, P_y, and P_z be the period lengths of generators X, Y, and Z, respectively. Denote the least common multiple and the greatest common divisor of P_x and P_y by $\mathrm{lcm}(P_x, P_y)$ and $\gcd(P_x, P_y)$, respectively.

Theoretical lower and upper bounds of P_z were provided in Li et al. (2012) as stated below in Theorem 5.8.

Theorem 5.8
$$\frac{\mathrm{lcm}(P_x, P_y)}{\gcd(P_x, P_y)} \leq P_z \leq \mathrm{lcm}(P_x, P_y). \tag{5.6}$$

The combination generator Z is likely (but not always) to have a period length that can reach the upper bound $\mathrm{lcm}(P_x, P_y)$. However, certain combinations may reduce its period length. For example, let C_i's be the sequence generated by a generator of a period length

P_c much shorter than P_x and P_y and let $Y_i = m - X_i + C_i$; then the combination generator becomes $Z_i = (X_i + Y_i) \bmod m$, which equals C_i, and thus has a period of length P_c. In particular, if we choose $Y_i = m - X_i + c$ for a constant $c \in \mathbb{Z}_m$, then $Z_i = c$, yielding a period of length 1. The latter example also shows that the lower bound given in Theorem 5.8 is a tight (i.e., reachable) lower bound. Generally speaking, the upper bound of $\text{lcm}(P_x, P_y)$ can be reached if the generating methods for X_i's and Y_i's are "unrelated". From statistical theory, X_i and Y_i being two statistically independent random variables would rule out the "bad" cases of $Y_i = m - X_i + C_i$. In fact, the upper bound is reached when $\gcd(P_x, P_y) = 1$, for which $P_z = P_x \times P_y$ by Theorem 5.8. This is a well-known result given in Knuth (1998).

From the predictability aspect, if we are given only the value of Z_i, i.e., $(X_i + Y_i) \bmod m$, one cannot easily separate the values of X_i and Y_i. We formalize this fact in the following theorem.

Theorem 5.9 *Let X and Y be two independent random variables representing two random number generators. If X and Y are uniformly distributed over $\mathbb{Z}_m$ and let $Z = X + Y \bmod m$, then*

1. *Z is uniformly distributed over $\mathbb{Z}_m$. (This result still holds even when only X or Y is uniformly distributed over $\mathbb{Z}_m$.)*
2. *For any $z \in \mathbb{Z}_m$, the conditional distributions of $X|Z = z$ and $Y|Z = z$ are uniformly distributed over $\mathbb{Z}_m$.*

The above two theorems provide some security justifications for the combination method. If we are given only the value of Z, according to Theorem 5.9, the value of X or Y is equally likely to be any value in $\mathbb{Z}_m$. That is, given any $z \in \mathbb{Z}_m$, $P(X = x|Z = z) = 1/m$ for all $x \in \mathbb{Z}_m$ and $P(Y = y|Z = z) = 1/m$ for all $y \in \mathbb{Z}_m$. At each iteration of the generating process, the combination method produces one final output variate (Z_i) from the two internal variates (X_i and Y_i). Then, by carefully choosing generators X and Y, the combination generator $Z = (X + Y) \bmod m$ would have a longer period (by Theorem 5.8) and better distributional property and unpredictability (by Theorem 5.9) than X and Y.

Theorem 5.9 justifies the ideal (but unrealistic) assumption that both generators have the exact uniform distributions over $\mathbb{Z}_m$. As argued in Marsaglia (1985) and Deng and George (1990), the distributional property of the combination generator can be improved even when the two baseline generators are not uniformly distributed.

Furthermore, from Theorem 5.9, given only the value of Z_i, it is hard to recover X_i or Y_i. However, we should mention that adding two linear generators yields another linear generator; see, for example, L'Ecuyer (1996). Consequently, if we are given enough consecutive variates of the combination generator, it is easy to break the system by solving a set of linear equations. To avoid that, it is suggested to include at least one non-linear generator to form a combination generator; for example, $Z = (f_1(X) + f_2(Y)) \bmod m$, where $f_1(\cdot)$ and $f_2(\cdot)$ are some efficient non-linear transformations discussed earlier.

To improve the generating efficiency of $Z = (X + Y) \bmod m$, one can choose the modulus $m = 2^w$, where w is a positive integer. Then, $Z = (X + Y) \bmod 2^w$ can be implemented as $(X + Y) \& (2^w - 1)$, where "&" is the logical "and" operation and $2^w - 1$ in its binary representation has w 1's in the w least significant bits and 0's in the rest of bits. The size of w in $m = 2^w$ is commonly chosen to be the CPU word size. In this paper, we fix $w = 32$, yet it can be easily increased to $w = 64$ or $w = 128$ for advanced CPUs, or decreased to $w = 16$ or $w = 8$ for simple CPUs.

Part II
Quality Assessment for Random Number Generators

Spectral Test for LCG/MRG

6

With increasing computing power and decreasing computing cost, it is common to perform a large-scale simulation study to evaluate the performance of newly proposed statistical methods, particularly when the analytical properties of a statistical procedure are intractable. To produce a sequence of random variates for a given distribution, most (if not all) algorithms are based on the assumption that one can produce a sequence of "independent" variates from the uniform distribution, U(0,1). However, no random number generator could claim to be "perfect" in the sense that it could produce a truly independent sequence with the exact uniform distribution. Clearly, the validity of such a large-scale computer simulation relies heavily on the "goodness" of the pseudo-random number generators used. Therefore, it is important to consider the issue of finding "better" random number generators.

Until recently, the linear congruential generators (LCGs), first proposed by Lehmer (1951), were the most popular. There are several reasons for the popularity of LCGs: (1) They are simple and efficient to implement because they are based on a simple first-order recurrence equation to produce the next variate; (2) Their theoretical properties have been studied extensively, including the period length and lattice structure. There are also several popular methods to construct improved generators over LCGs.

Example 6.1 (*Spectral test for ranking LCGs*)

Knuth (1998) discussed extensively the spectral test for LCGs, which is a theoretical test that provides a measure of uniformity in higher dimensions. Therefore, one can use the spectral test in Knuth (1998) to rank the "goodness" of various LCGs.

L. Deng et al., *Random Number Generators for Computer Simulation and Cyber Security*, Synthesis Lectures on Mathematics & Statistics, https://doi.org/10.1007/978-3-031-76722-7_6

Example 6.2 (*Wichmann and Hill generators*)

Wichmann and Hill (1982) proposed combining three different LCGs to form a generator that has a longer period length and better spectral properties over the individual component LCGs.

Wichmann and Hill generator became quite popular but certain potential problems were also found: (i) McLeod (1985) pointed out that it may not produce variates strictly between 0 and 1, which may affect its portability property; (ii) Zeisel (1986) pointed out that it is equivalent to another LCG with a large multiplier and a large modulus; hence, Wichmann and Hill generator may share the same defect as other LCGs. Another example is the multiple recursive generator (MRG), a natural extension of the LCG, that uses a large-order recurrence equation to compute its next variate. Recently, LCGs have been replaced by large-order maximum-period multiple MRGs and other generators (e.g., MT19937) in the area of computer simulation. They have a long period and the nice property of high-dimensional equi-distribution. Several classes of large-order MRGs have been proposed in the literature in recent years (see, e.g., Deng and Shiau 2015).

The spectral test is a theoretical test that provides a measure of uniformity in dimensions beyond the order (k) of the MRG. It is often used to rank MRGs of the same order. To the best of our knowledge, no efficient algorithms are available in the literature for computing the spectral test of large-order MRGs.

In this chapter, we give an introduction to the spectral test and show how it is computed for LCGs. Then, in the next chapter, we will present how the spectral test is computed for MRGs, including a recent more efficient method compared to the prevalent one.

Notations

For a prime p, following L'Ecuyer and Simard (2014), we define, for any integer x,

$$[x]_p = \begin{cases} (x \bmod p), & \text{if } (x \bmod p) < p/2; \\ (x \bmod p) - p, & \text{otherwise.} \end{cases}$$

Therefore, $[x]_p$ denotes the symmetric representation of x with respect to modulus p such that $-p/2 < [x]_p < p/2$ for every odd prime p. In what follows we choose p to be a large prime, so it is odd, and we use this inequality freely. Notice that $[x]_p$ is the number smallest in absolute value in the congruence class modulo p of x. For a t-dimensional integer vector $\mathbf{x} = (x_1, x_2, \ldots, x_t)'$, we define $[\mathbf{x}]_p = ([x_1]_p, [x_2]_p, \ldots, [x_t]_p)'$ and $\|\mathbf{x}\|^2 = \sum_{i=1}^t x_i^2$.

In this chapter, in addition to the regular modulo function for the computation of the next generated number of an MRG, we will need to apply the symmetric modulo function $[x]_p$ to some integers x to solve certain minimization problems. Because of this, the MRG multipliers $\alpha_1, \alpha_2, \ldots, \alpha_k$ and some "constant multiplier" c will be expressed in "symmetric-modulus" format with their absolute values less than $p/2$.

We briefly recall concepts introduced earlier. An MRG generates pseudo-random numbers sequentially with a k-th order linear recurrence:

$$X_i = (\alpha_1 X_{i-1} + \alpha_2 X_{i-2} + \cdots + \alpha_k X_{i-k}) \bmod p, \quad i \geq k, \tag{6.1}$$

where multipliers $\alpha_1, \alpha_2, \ldots, \alpha_k$ are integers in $\mathbb{Z}_p$, $\alpha_k \neq 0$, and one can choose any k not-all-zero integers $X_0, X_1, \ldots, X_{k-1}$ to be the starting seeds. It is well known that the MRG in Eq. 6.1 has the maximum period, $p^k - 1$, if and only if its characteristic polynomial

$$f(x) = x^k - \alpha_1 x^{k-1} - \alpha_2 x^{k-2} - \cdots - \alpha_k$$

is a k-th degree primitive polynomial over $\mathbb{Z}_p$. To convert the variates strictly between 0 and 1, we can use $U_i = (X_i + 1/2)/p$. For the remainder of the chapter, we will only consider maximum-period MRGs. We remark that when the order $k = 1$, the MRG reduces to the LCG: linear congruential generator (LCG) proposed by Lehmer (1951):

$$X_i = B X_{i-1} \bmod p, \quad i \geq 0. \tag{6.2}$$

6.1   Equi-Distribution Property and the Spectral Test

We discuss the spectral test for both LCGs and MRGs, as such we first recall the MRGs briefly.

It is well known that maximum-period MRGs have the equi-distribution property up to order k; that is, over its entire period of $p^k - 1$, every t-tuple $(1 \leq t \leq k)$ of integers in $\mathbb{Z}_p^t$ appears exactly the same number of times (p^{k-t}), with the exception of the all-zero tuple that appears one time less (see, e.g., Lidl and Niederreiter 1994, Theorem 7.43). On average, a true t-dimensional multivariate uniform distribution would produce each of p^t t-tuples in $\mathbb{Z}_p^t$ with equal frequency for any dimension t. Therefore, the large period length and high-dimensional equi-distribution properties become even more advantageous as k gets larger. In fact, for $t \leq k$, large-order MRGs are pretty close to an "ideal" generator—only the all-zero tuple is generated one less time than the other t-tuples.

To compute the spectral test of an MRG, we consider all the successive sequences of length t generated from the Eq. 6.1: $\mathbf{S}_n = (X_n, X_{n+1}, \ldots, X_{n+t-1})'$ for $n = 0, 1, \ldots, \rho - 1$, where $\rho = p^k - 1$ is the period length. $\mathbf{S}_n$ is referred to as the state of the MRG at step n. Let I be a set of fixed nonnegative integers. The particular set of integers $I = \{0, 1, \ldots, t - 1\}$ could be thought of as the indices of the state $\mathbf{S}_n$, which create all the possible successive t-tuples over all steps, $n = 0, 1, \ldots, \rho - 1$, in the entire period of the MRG.

As discussed in L'Ecuyer and Simard (2014), the index set I is not required to be a set of successive integers; more generally, it can be any set of non-negative integers $I =$

$\{j_1, j_2, \ldots, j_r\}$, where $j_1 < j_2 < \cdots < j_r$ and r is the number of indices in I. Let $t = j_r - j_1 + 1$, the number of variates that need to be generated at each step of the MRG state $\mathbf{S}_n$. Denote the set of all possible r-tuples that a maximum-period MRG can generate according to I by

$$L_r(I) = \left\{ (X_{n+j_1}, X_{n+j_2}, \ldots, X_{n+j_r}) | n = 0, 1, \ldots, \rho - 1 \right\}. \tag{6.3}$$

According to the equi-distribution property, for *any* choice of I with $j_r - j_1 < k$, or equivalently, $t(= j_r - j_1 + 1) \leq k$, every nonzero r-tuple will appear the same number of times in $L_r(I)$. However, when $j_r - j_1 \geq k$ (i.e., $t > k$), the equi-distribution property is impossible to achieve. This can be clearly seen by the following argument. Ideally, we would want to generate all p^t possible t-tuples in $\mathbb{Z}_p^t$ with equal frequency over the entire period of the MRG and then choose r elements from each of these t-tuples in accordance with the index set I. However, because the MRG can only generate $p^k - 1$ numbers before the sequence starts to repeat, many t-tuples would never be generated, which implies many r-tuples would never be produced.

Geometrically, $L_r(I)$, the set of r-tuples that can be generated, forms a lattice of points in the r-dimensional space $\mathbb{Z}_p^r$. A lattice of points in $[0, 1)^r$, $\Lambda_r(I)$, can be obtained by scaling each r-tuple in $L_r(I)$ by p, i.e.,

$$\Lambda_r(I) = \left\{ \left(\frac{X_{n+j_1}}{p}, \frac{X_{n+j_2}}{p}, \ldots, \frac{X_{n+j_r}}{p} \right) \Big| n = 0, 1, \ldots, \rho - 1 \right\}. \tag{6.4}$$

When $t(= j_r - j_1 + 1) > k$, like the well-known problem for the LCG (Marsaglia 1968), these r-dimensional points form a lattice in an r-dimensional hypercube where we can find families of equidistant parallel $(r - 1)$-dimensional hyperplanes to cover all the points in the lattice. The lattice points in the spacing between parallel hyperplanes are the r-tuples that the MRG could never produce; thus, the smaller the spacing, the better the MRG.

The spectral test computes the largest distance between adjacent parallel hyperplanes among families of parallel hyperplanes that cover all the points in $\Lambda_r(I)$, or equivalently, $L_r(I)$ (see, e.g., Knuth 1998, L'Ecuyer 1997). This largest distance is referred to as the *spectral distance* in the literature and is often denoted by $d_r(k)$. The spectral distance is a measure of the uniform spread of the r-tuples across the r-dimensional space, and a small spectral distance implies a more uniform coverage. Therefore, a large $d_r(k)$ is considered "bad," for it would take only a relatively small number of parallel $(r - 1)$-dimensional hyperplanes to cover all the r-dimensional lattice points produced by the generator; and, consequently, the generator is often said to have a bad lattice structure in dimension r because of the large distance between hyperplanes. Clearly, when $t(= j_r - j_1 + 1)$ is much larger than k, regardless of how small r is, the spectral distance $d_r(k)$ becomes so large that no MRGs (of fixed order k) can be considered "good" when evaluated by $d_r(k)$.

The spectral distance is classically computed for the set of points using successive indices in an index set $I = \{0, 1, \ldots, t - 1\}$ (see, e.g., Knuth 1998) as mentioned above. There are other ways to generate such points (for example using non-successive indices) but for the remainder of this chapter, we will only consider the classical spectral test with t successive indices and compute the spectral distance $d_t(k)$ accordingly.

In the following subsections, we will describe a simple and intuitive geometric interpretation of the spectral test as a minimization problem of finding the normal vector (i.e., the vector perpendicular to the parallel hyperplanes covering all the points in $\Lambda_t(I)$) with the shortest Euclidean length. Then in Sect. 7.2.1, we will outline a new method for computing the spectral test.

6.2 Lattice Structure

For MRG of order k, when $t > k$, the number of possible t-tuples is p^t which is larger than the MRG's period $p^k - 1$. Therefore, there will be many t-tuples, or successive sequences, that the MRG *cannot* generate. A plot of all possible t-tuples that the MRG can generate will reveal a *lattice* of points in a t-dimensional hypercube and a relatively small number of green families of equidistant parallel hyperplanes can cover all points.

For LCGs and low-order MRGs, it is common to evaluate their performance using the lattice structure criterion. More specifically, we study the structure of the d consecutive elements of the sequence, $\{(X_i, X_{i+1}, \ldots, X_{i+d-1}) \mid i = 0, 1, \ldots\}$, produced by a random number generator, or equivalently by its uniform$(0,1)$ counterpart, $\{(U_i, U_{i+1}, \ldots, U_{i+d-1}) \mid i = 0, 1, \ldots\}$. If the generated sequence is indeed a realization of a sequence of truly independent uniform random variables, then these d-tuples should be uniformly distributed over the d-dimensional cube. For LCGs, Marsaglia (1968) was the first to show that successive overlapping sequences of d random numbers fall on at most $(d\,!\,m)^{1/d}$ hyperplanes, where m is the modulus chosen. This shortcoming may yield grossly wrong results for certain applications, such as in the Monte Carlo multiple-integration method. For MRGs, when $d > k$, all the d consecutive points lie on some parallel hyperplanes in the d-dimensional space and its d-dimensional lattice structure can determine the property of the MRG.

One quantitative measure of the lattice structure is the spectral test corresponding to the maximum-distance between two adjacent parallel hyperplanes. Clearly, we would prefer the generator with the smaller maximum-distance because no points in between these adjacent hyperplanes can be generated. In theory, a good uniform random number generator should produce points that fill evenly over the whole space. A smaller maximum-distance may avoid large slices of empty space so that the generated number sequences can be more uniformly distributed over the whole space. L'Ecuyer (1997) pointed out that a necessary but not sufficient condition for an MRG to have a good lattice structure is that the sum of squares of all coefficients, $\sum_{i=1}^{k} \alpha_i^2$, is large. A similar conclusion has been made in Deng et al. (1997) from a statistical justification viewpoint. Consequently, for DX-k-s generators, this means that we prefer a large order k and large values of s and B.

Spectral test is considered as a "gold standard" to evaluate the "goodness" of linear generators like LCGs or MRGs. The most popular algorithm proposed can be efficient for LCGs but not for large-order MRGs. To motivate and compare the recently proposed efficient algorithm with the classical algorithm, we will start by describing the issue of spectral tests for LCGs.

6.3   Computing the Spectral Test for LCGs

We look at the relatively simple case of an LCG $X_0, X_1, \ldots$, given by the equation,

$$X_i = B X_{i-1} \bmod p,$$

where we may assume B is a primitive root modulo prime p (as such this LCG is also a maximum period LCG), and X_0 is the initial seed. We may assume that $X_0 \neq 0$. It is easy to see from here that, $X_i = B^i X_0 \bmod p$ and if we want to compute the spectral test in $t = 2$ dimensions we will look at the normalized successive 2-tuples $(X_i/p, X_{i+1}/p) = (\frac{B^i X_0}{p}, \frac{B^{i+1} X_0}{p})$ where we assume we are working modulo p when computing the $B^i X_0, B^{i+1} X_0$. These points are all in the unit square $[0, 1] \times [0, 1]$. The spectral distance test is determined by the maximum distance between parallel hyperplanes (lines in this case) required to cover these points. This distance is not affected if the region is extended to the entire plane by just shifting. So the unit square for example would need to be shifted by all integral distances in both dimensions. The advantage of this is that the resulting set of points is a lattice which enables the use of algorithms for lattices. Moreover, the modulo p operation can also be "hidden" by introducing an additive integer. So, we could be considering all points of the form $(\frac{B^i X_0}{p} + k_1, \frac{B^{i+1} X_0}{p} + k_2)$ where k_1, k_2 are arbitrary integers. We may assume that $k_1 = 0$ since the $B^i X_0$ can be an arbitrary integer x (in $[1, p-1]$ as i varies) so it can also take the value $B^i X_0 + k_1 p$ as well, thus effectively k_1 can be assumed to be 0. This does require a slight adjustment because we want to still have the second coordinate as B times the first one, and if $x \to x + k_1 p$ then the second coordinate is now not exactly B times the first one plus $k_2 p$. But this is okay, as we can assume $k_2 \to k_2 + B k_1$ and this is allowed as k_2 is an arbitrary integer. Thus the set of points is more simply put as, $(\frac{x}{p}, \frac{Bx}{p} + k_2)$ where x is an arbitrary integer in $[1, p-1]$. To this set we need to add the point $(0, 0)$ as that is not generated, but at the same time, does not affect the results. The final set of points can be represented as the set,

$$\Lambda_2(I) = \{y_1 \mathbf{v}_1 + y_2 \mathbf{v}_2 : y_1, y_2 \text{ integer }\},$$

where $\mathbf{v}_1 = (\frac{1}{p}, \frac{B}{p})$, $\mathbf{v}_2 = (0, 1)$ (so for generating $(\frac{x}{p}, \frac{Bx}{p} + k_2)$ we will take $y_1 = x, y_2 = k_2$.) These vectors $\mathbf{v}_1, \mathbf{v}_2$ are said to generate the lattice in question and are a basis for the lattice. Here is an example that illustrates this construction:

Example 6.3 (*LCG lattice*)

Let $p = 23$, $B = 5$. It can be checked that B is a primitive root modulo $p = 23$ since $B^{11} \equiv -1 \bmod 23$. Now, let the initial seed be $X_0 = 2$. The generated sequence is,

$$2, 10, 4, 20, 8, 17, 16, 11, 9, 22, 18, 21, 13, 19, 3, 15, 6, 7, 12, 14, 1, 5, 2, \ldots,$$

If we consider all the successive tuples formed they are,

$$(2, 10), (10, 4), (4, 20), (20, 8), (8, 17), (17, 16), (16, 11), (11, 9), (9, 22),$$
$$(22, 18), (18, 21), (21, 13), (13, 19), (19, 3), (3, 15), (15, 6), (6, 7), (7, 12),$$
$$(12, 14), (14, 1), (1, 5), (5, 2).$$

We add $(0, 0)$ to the above set of tuples, and divide each coordinate by $p = 23$. The resulting set of points lies in the unit square as shown

---

Which set of parallel equi-spaced lines will give the maximum spacing, if they cover all these points? This seems a difficult problem, but fortunately, it can be converted to a question about the shortest vector in the *dual lattice* of the above one. The algorithm is described in Knuth (1998). Below, we outline a simpler method that is essentially the same as this method but *works off a different basis* and, for this basis, the dual basis is easier to see and work with. For the case of 2 dimensions and an LCG, computing the basis as we just did, does not seem tough. However, this difference is huge when we are working with MRG's of order k much larger than one and we want to compute the spectral test in $t = k + d$ dimensions. Below, we continue explaining and motivating our method explained in the next chapter, using the LCG as an example.

We start with computing the spectral distance $d_2(1)$ for an LCG as in Eq. 6.2, which is also an MRG of order $k = 1$. The maximum period of an LCG is $\rho = p - 1$. Let $I = \{0, 1\}$ and $\Lambda_2(I)$ as in Eq. 6.4 be the set of all successive (overlapping) pairs $(X_n/p, X_{n+1}/p)$ for $n = 0, 1, \ldots, \rho - 1$. Clearly, the ordered pairs in $\Lambda_2(I)$ can be covered by several parallel lines of the form $Bx - y = e$, where x corresponds to X_n/p, y corresponds to X_{n+1}/p, and e is the integer multiple of p such that $BX_n - X_{n+1} = ep$. The normal vector to these parallel lines is $\mathbf{w} = (B, -1)'$. It can be easily shown that the distance between any two adjacent parallel lines is $1/\|\mathbf{w}\| = 1/\sqrt{1 + B^2}$; hence, a shorter $\|\mathbf{w}\|$ implies a larger distance between a pair of adjacent parallel lines.

For an integer c, every generated output pair in $L_2(I)$ also satisfies the equation

$$cX_i = [cB]_p X_{i-1} \bmod p.$$

Hence, the points in lattice $\Lambda_2(I)$ can also be covered by parallel lines of the form $[cB]_p x - cy = e$, of which the corresponding normal vector is $\mathbf{N}_c = [c\mathbf{w}]_p = ([cB]_p, -c)'$. Then the distance between two adjacent parallel lines is $1/\|\mathbf{N}_c\| = 1/\sqrt{c^2 + [cB]_p^2}$. It is easy to see that $|| [-c\mathbf{w}]_p || = || [c\mathbf{w}]_p ||$; hence, we can restrict the range of c to $0 < c < p/2$. Each c in this range defines a (not necessarily unique) family of parallel lines that cover all the points in $\Lambda_2(I)$ with the distance between adjacent parallel lines being $1/||\mathbf{N}_c||$. Therefore, the spectral distance $d_t(k)$ is equal to the largest $1/||\mathbf{N}_c||$ among all c in $0 < c < p/2$.

For $p = 31$, the questions are: how many Linear Congruential Generators (LCGs) exist, and how can they be found or generated? There are straightforward mathematical formulas and methods for this, and there are eight primitive roots: 3, 11, 12, 13, 17, 21, 22, 24. The next question is, which one is the best? What criteria should be used to determine this?

Example 6.4 (*LCG with p = 31*)

We consider an LCG with multiplier $B = 3$ and modulus $p = 31$, which we denote by LCG(3, 31). In Fig. 6.1, we plot several families of parallel lines that cover all the points in $\Lambda_2(I)$, the successive overlapping pairs generated from LCG(3, 31). The normal vector $\mathbf{N}_c = [c\mathbf{w}]_p$ of each family is given in the caption of each subfigure. Note that the cases of $c = 2, 3, 4, 5$ are not shown because their plots are the same as that of $c = 1$. Also, the cases of c in the range of negative range $(-p/2 < c < 0)$ are omitted for they duplicate the c's in the range of $0 < c < p/2$. Because the normal vector $\mathbf{N}_1 = [1\mathbf{w}]_p = (3, -1)'$ is of the shortest length, the family corresponding to $c = 1$ is the one with the largest spectral distance between adjacent parallel lines, which can be clearly seen from Fig. 6.1.

Example 6.5 (*LCG(B = 12, p = 31)*)

Similar to Example 4, the spectral distance can be calculated for LCG(B = 12, p = 31). Figure 6.2 show the distances based on various c's. It is shown that $c = 5$ is the one with the largest spectral distance between adjacent parallel lines.

Example 6.6 (*Spectral test of LCGs with p 31*)

The method for determining the maximal spectral distance can be applied to all values of B for LCG(B, $p = 31$). Given that the primitive roots are 3, 11, 12, 13, 17, 21, 22, and 24, one can evaluate $B = 3, 11, 12, 13, 17, 21, 22, 24$, and select the value with the smallest maximal distance. Table 6.1 shows that when $B = 12$ or 13, the LCGs have the smallest maximal distance. More details are shown in Fig. 6.3.

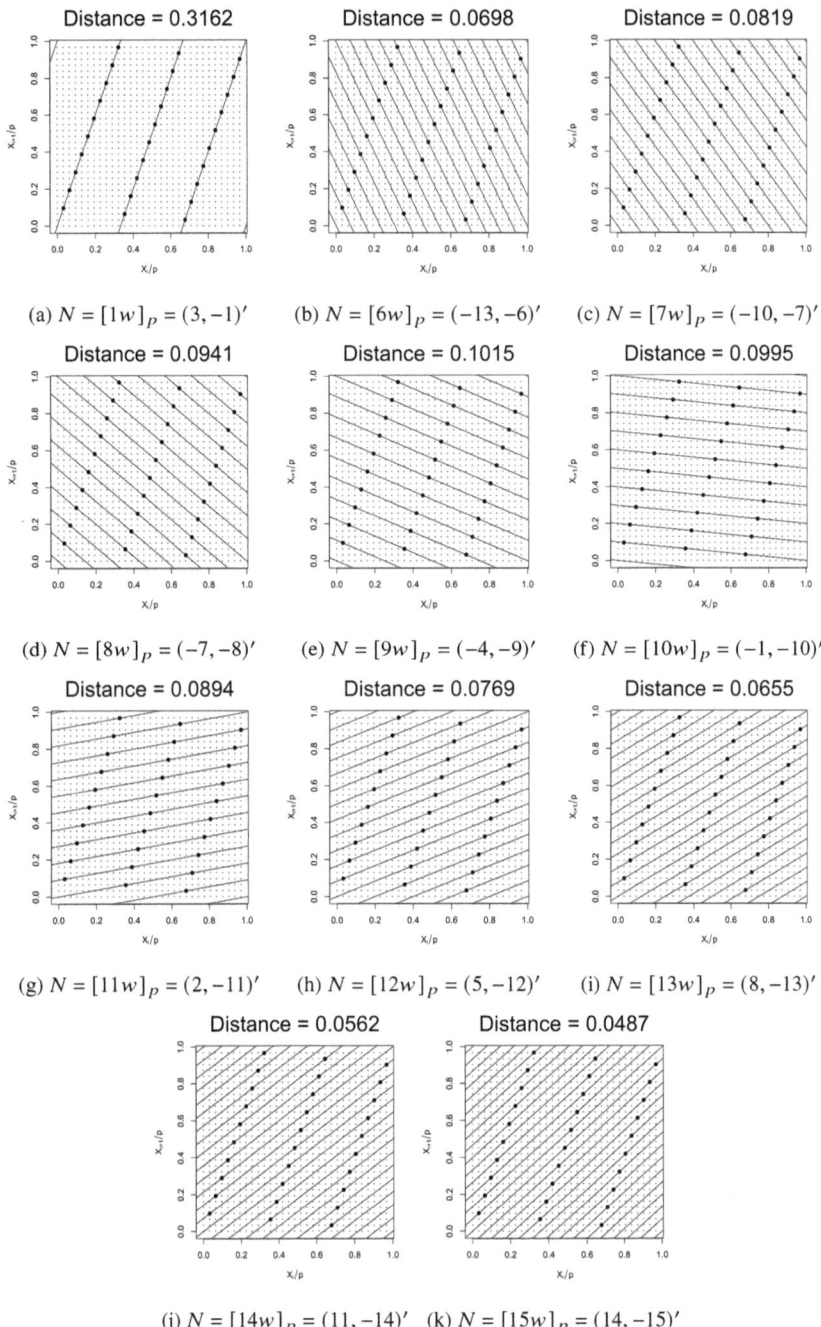

Fig. 6.1 Distinct families of parallel lines covering successive overlapping pairs from LCG(3, 31)

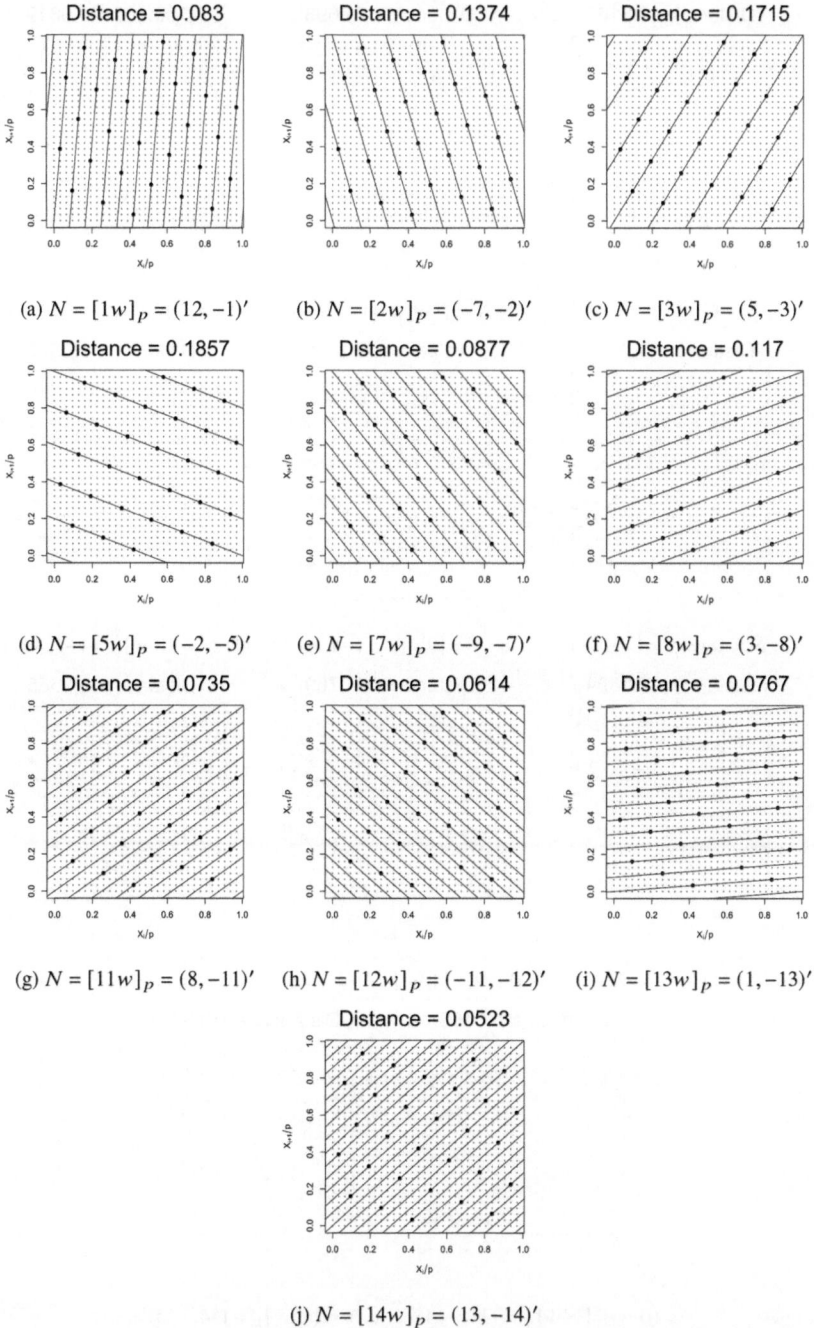

Fig. 6.2 Distinct families of parallel lines covering successive overlapping pairs from LCG(12, 31)

Table 6.1 The maximum separation between the parallel lines when various values of B are chosen. Specifically, when B is set to 12 or 13, the maximum separation between the parallel lines is minimized, which is the preferred outcome

B	c	N	Spectral distance
12	5	$(-2, -5)$	0.1857
13	2	$(-5, -2)$	0.1857
22	3	$(4, -3)$	0.2000
24	4	$(3, -4)$	0.2000
11	3	$(2, -3)$	0.2774
17	2	$(3, -2)$	0.2774
3	1	$(3, -1)$	0.3162
21	3	$(1, -3)$	0.3162

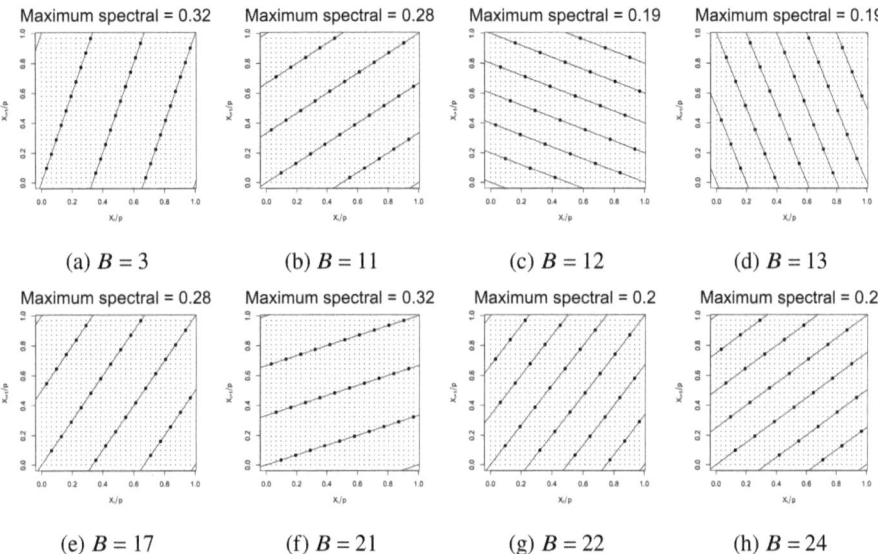

Fig. 6.3 The maximum separation between the parallel lines when various values of B are chosen. Specifically, when B is set to 12 or 13, the maximum separation between the parallel lines (as shown in solid lines) is minimized, which is the preferred outcome

From the previous examples, for $t = 2$ dimensions, the "*best*" LCGs are those with $B = 12$ and $B = 13$, whereas the "*worst*" LCGs are those with $B = 3$ and $B = 21$. The next natural questions are: *What about t > 2 dimensions?* and *How to compute the spectral distance?*

6.4   Classical LCG Spectral Test in Higher Dimensions

We first describe the classical algorithm as given in Knuth (1998).

Algorithm 1: Algorithm S (Knuth 1998)

For a given $B \in \mathbb{Z}_p$,

$$v_t = \min_{(x_1,\dots,x_t)\in\mathbb{Z}_p^t} \left\{ \sqrt{x_1^2 + \cdots + x_t^2} \,\middle|\, x_1 + Bx_2 + \cdots + B^{t-1}x_t = 0 \bmod p \right\}$$

where $2 \le t \le T$ with some specified dimensions T.

The Algorithm S is somewhat intricate, and the theory behind the test was presented in Knuth (1998). Once v_t has been computed, $d_t = 1/v_t$ represents the maximum distance between t-dimensional hyperplanes, considering all families of parallel hyperplanes covering the points in t dimensions. Furthermore, extending this method to the spectral test of MRG is not straightforward. There is a demand for an efficient algorithm applicable to both LCG and MRG.

6.5   More Efficient LCG Spectral Test in Higher Dimensions

The key difference between the classical algorithm and the proposed algorithm (which will be presented in the next chapter) is the choice of "normal vectors". Consider a 5-dimensional hyperplane (t = 5) as an example:

1. Algorithm S: the set of normal vectors for the 5-dimensional plane is

$$\mathbf{v}1 = [B, -1, 0, 0, 0]'$$
$$\mathbf{v}2 = [B^2 \bmod p, 0, -1, 0, 0]'$$
$$\mathbf{v}3 = [B^3 \bmod p, 0, 0, -1, 0]'$$
$$\mathbf{v}4 = [B^4 \bmod p, 0, 0, 0, -1]'$$

2. New method: the set of normal vectors for the 5-dimensional plane is

$$\mathbf{v}1 = [B, -1, 0, 0, 0]'$$
$$\mathbf{v}2 = [0, B, -1, 0, 0]'$$
$$\mathbf{v}3 = [0, 0, B, -1, 0]'$$
$$\mathbf{v}4 = [0, 0, 0, B, -1]'$$

Table 6.2 Spectral test of LCG(B, p = 31) with with $t = 3$ dimensions

B	c_1	c_2	N	Spectral
3	1	0	$(3, -1, 0)$	0.3162
11	0	1	$(-3, 0, -1)$	0.3162
17	0	3	$(-1, 0, -3)$	0.3162
21	1	-3	$(0, -1, 3)$	0.3162
22	2	1	$(1, -2, -1)$	0.4082
24	2	-1	$(-1, -2, 1)$	0.4082
12	1	1	$(1, -1, -1)$	0.5774
13	1	-1	$(-1, -1, 1)$	0.5774

Once the normal vectors are selected, we can compute the spectral test value as solving the minimization problem

$$v_5^2 = \min_{(c_1, c_2, c_3, c_4) \neq (0,0,0,0)} \left(|| \sum_{i=1}^{4} [c_i \mathbf{v}i]_p ||^2 \right)$$

Note that the classical Algorithm S requires the computation of normal vectors, whereas the new algorithm can find them quickly without any computation. For LCG, the difference between the algorithms is not significant. However, for large-order MRGs, it is clear that the new algorithm is much more intuitive and efficient.

Example 6.7 (*Spectral test of LCG(B, p = 31) with t = 3 dimensions*)

For $t = 3$ dimensions, Table 6.2 displays the maximal spectral distance of the LCG(B, $p = 31$) for different values of B.

The "best" LCGs are $B = 3$, $B = 11$, $B = 17$, and $B = 21$, while the "worst" LCGs are $B = 12$ and $B = 13$. However, recall that for $t = 2$ dimensions, the "best" LCGs are $B = 12$ and $B = 13$, while the "worst" LCGs are $B = 3$ and $B = 21$. Therefore, caution is needed when ranking different LCGs.

Efficient Implementation for Spectral Test

We first describe the existing prevalent method for the computation of the spectral test, followed by a recent more efficient one.

7.1 Existing Computational Method for Spectral Test

We now describe an existing method for computing the spectral test that is similar to the ones used in Kao and Tang (1997), Knuth (1998), L'Ecuyer (1997), L'Ecuyer and Couture (1997). First, let $\mathbf{A}$ be the transpose of the MRG's companion matrix, i.e.,

$$\mathbf{A} = \begin{bmatrix} 0 & 0 & \dots & 0 & \alpha_k \\ 1 & 0 & \dots & 0 & \alpha_{k-1} \\ 0 & 1 & \dots & 0 & \alpha_{k-2} \\ \vdots & \vdots & \ddots & \vdots & \vdots \\ 0 & 0 & \dots & 1 & \alpha_1 \end{bmatrix},$$

and denote the ijth element of $\mathbf{A}$ by $[\mathbf{A}]_{ij}$. The commonly used method defines a basis matrix for $L_t(I) + p\mathbb{Z}^t$ as

$$\mathbf{V} = \begin{bmatrix} \mathbf{I}_{k \times k} & \mathbf{X}_{k \times d} \\ \mathbf{0}_{d \times k} & p\mathbf{I}_{d \times d} \end{bmatrix},$$

where $\mathbf{I}_{k \times k}$ is the identity matrix, $\mathbf{0}_{d \times k}$ is the zero matrix, and

© The Author(s), under exclusive license to Springer Nature Switzerland AG 2025
L. Deng et al., *Random Number Generators for Computer Simulation and Cyber Security*, Synthesis Lectures on Mathematics & Statistics,
https://doi.org/10.1007/978-3-031-76722-7_7

$$\mathbf{X}_{k \times d} = \begin{bmatrix} \alpha_k & [\mathbf{A}^2]_{1k} \bmod p & \cdots & [\mathbf{A}^d]_{1k} \bmod p \\ \alpha_{k-1} & [\mathbf{A}^2]_{2k} \bmod p & \cdots & [\mathbf{A}^d]_{2k} \bmod p \\ \vdots & \vdots & \ddots & \vdots \\ \alpha_1 & [\mathbf{A}^2]_{kk} \bmod p & \cdots & [\mathbf{A}^d]_{kk} \bmod p \end{bmatrix}.$$

Before we go on further, it is worthwhile to pause a while to understand roughly why the above matrix provides a basis for the lattice, at a level similar to what we did for LCG's in Sect. 6.3. This will help us better appreciate the more efficient method described later. Firstly, here to make things easier we are not working on the normalized MRG sequence, but on the original one $X_0, X_1, \ldots, X_{k-1}, X_k, \ldots$, where we can assume $X_0, X_1, \ldots, X_{k-1}$ are the seeds and $X_i = \alpha_1 X_{i-1} + \alpha_2 X_{i-2} + \cdots + \alpha_k X_{i-k} \bmod p$ for $i \geq k$. Next, to simplify things we assume we are working modulo p so we can drop the $\bmod p$ for brevity. We also assume that the MRG is full period. Consider one of the successive $t = (k + d)$ tuples $(X_n, X_{n+1}, \ldots, X_{n+t-1})$. This is a point in $[0, p-1]^t$, but since we are extending the set of points by translating the cube $[0, p-1]^t$ by multiples of p in all directions, a general point in the overall lattice may be written as, $(X_n + m_0 p, X_{n+1} + m_1 p, \ldots, X_{n+t-1} + m_{t-1} p)$, where $m_0, \ldots, m_{t-1}$ are integers. Our goal is to be able to view these set of points as the set of points generated by taking linear combinations $y_1 \mathbf{v}_1 + y_2 \mathbf{v}_2 + \cdots + y_t \mathbf{v}_t$ for all possible integers $y_1, \ldots, y_t$ and some vectors $\mathbf{v}_1, \ldots, \mathbf{v}_t$. In other words we want to come up with a basis of the lattice.

An important point to notice here is that, similar to the LCG case (see Sect. 6.3) we may assume that $m_0, \ldots, m_{k-1}$ are all zero. This is so because since the MRG is full period, the $X_n, X_{n+1}, \ldots, X_{n+k-1}$ may be considered as "independent random variables" and all possible k-tuples from $[0, p-1]^k$ are possible (this is part of the equi-distribution property). Because of this, as n varies, and as $m_0, \ldots, m_{k-1}$ vary over all integers, the first k components vary over all the integers. We may thus take them to be arbitrary. This does complicate the matter a little bit though, since later as we will see we want to express the $X_{n+k}, \ldots, X_{n+t-1}$ as linear functions of $X_n, X_{n+1}, \ldots, X_{n+k-1}$ using the recurrence (this is derived below). But if now we are replacing X_n by say $X_n + m_0 p$, this will complicate things. But luckily there is an easy way out of this. Assuming, for a moment, $m_0 = m_1 = \ldots = m_{k-1} = 0$ we can express X_k as a linear function of the $X_0, X_1, \ldots, X_{k-1}$, say $X_k = F_k(X_0, \ldots, X_{k-1})$. Then, since, $F_k(X_0 + m_0 p, X_1 + m_1 p, \ldots, X_{k-1} + m_{k-1} p) = F_k(X_0, \ldots, X_{k-1}) + F_k(m_0 p, \ldots, m_{k-1} p)$ we can "adjust" this in the $m_k p$ by letting $m_k \to m_k + F_k(m_0 p, \ldots, m_{k-1} p)/p$. Since the m_k is arbitrary this is allowed. In other words, it is legitimate to consider $m_0 = m_1 = \cdots = m_{k-1} = 0$ and to avoid this re-adjustment of $m_k, m_{k+1}, \ldots,$ etc. we shall assume below that $m_0 = m_1 = \cdots = m_{k-1} = 0$, remembering that in reality there is an adjustment that can be made to account for this assumption.

Now, for the remaining coordinates of the arbitrary point $(X_n, X_{n+1}, \ldots, X_{n+k-1}, X_{n+k} + m_k p, \ldots, X_{n+t-1} + m_{t-1} p)$ (notice we already put $m_0 = m_1 = \cdots = m_{k-1} = 0$) we can do some computations as follows. First $X_{n+k} = \alpha_1 X_{n+k-1} + \alpha_2 X_{n+k-2} + \cdots + \alpha_k X_n$. And, we can write,

$$
\begin{aligned}
X_{n+k+1} &= \alpha_1 X_{n+k} + \alpha_2 X_{n+k-1} + \cdots + \alpha_k X_{n+1} \\
&= \alpha_1 (\alpha_1 X_{n+k-1} + \alpha_2 X_{n+k-2} + \cdots + \alpha_k X_n) + \alpha_2 X_{n+k-1} + \cdots + \alpha_k X_{n+1} \\
&= (\alpha_1^2 + \alpha_2) X_{n+k-1} + (\alpha_1 \alpha_2 + \alpha_3) X_{n+k-2} + \cdots + (\alpha_1 \alpha_{k-1} + \alpha_k) X_{n+1} + \alpha_1 \alpha_k X_n
\end{aligned}
$$

If we keep doing these computations for $X_{n+k+2}, X_{n+k+3}, \ldots$, each time expressing them ultimately as combinations of $X_n, X_{n+1}, \ldots, X_{n+k-1}$ we get the coefficients from the matrix $\mathbf{V}$ described earlier, in the upper right hand corner (top k rows of the last d columns), i.e., the matrix $\mathbf{X}_{k \times d}$. For example, from the above the coefficient of X_{n+k-1} for X_{n+k+1} is $(\alpha_1^2 + \alpha_2)$ which may be verified to be $[\mathbf{A}^2]_{1k}$ (notice that the mod p operation needs to be done as well in the computations). What about the lower right hand corner matrix, the $p\mathbf{I}_{d \times d}$? These serve to account for the $m_k p, m_{k+1} p, \ldots, m_{t-1} p$ among the last d coordinates of the point $(X_n, X_{n+1}, \ldots, X_{n+k-1}, X_{n+k} + m_k p, \ldots, X_{n+t-1} + m_{t-1} p)$. So, in the linear combination equation:

$$
(X_n, X_{n+1}, \ldots, X_{n+k-1}, X_{n+k} + m_k p, \ldots, X_{n+t-1} + m_{t-1} p) = y_1 \mathbf{v}_1 + y_2 \mathbf{v}_2 + \cdots + y_t \mathbf{v}_t,
$$

we may take $y_1 = X_n, y_2 = X_{n+1}, \ldots, y_k = X_{n+k-1}, y_{k+1} = m_{k+1}, y_{k+2} = m_{k+2}, \ldots, y_t = m_t$, and the vectors $\mathbf{v}_1, \mathbf{v}_2, \ldots, \mathbf{v}_t$ as the rows of the matrix $\mathbf{V}$ above. Perhaps, a more slick way to see the powers of $\mathbf{A}$ arising is as follows. One can verify the matrix equation,

$$
\begin{bmatrix} X_{n+1} & X_{n+2} & \ldots & X_{n+k} \end{bmatrix} = \begin{bmatrix} X_n & X_{n+1} & \ldots & X_{n+k-1} \end{bmatrix} \mathbf{A},
$$

and similarly,

$$
\begin{bmatrix} X_{n+2} & X_{n+3} & \ldots & X_{n+k+1} \end{bmatrix} = \begin{bmatrix} X_{n+1} & X_{n+2} & \ldots & X_{n+k} \end{bmatrix} \mathbf{A} = \begin{bmatrix} X_n & X_{n+1} & \ldots & X_{n+k-1} \end{bmatrix} \mathbf{A}^2.
$$

Continuing in this way, we can show (for example by induction) that,

$$
\begin{bmatrix} X_{n+d} & X_{n+d+1} & \ldots & X_{n+d+k-1} \end{bmatrix} = \begin{bmatrix} X_n & X_{n+1} & \ldots & X_{n+k-1} \end{bmatrix} \mathbf{A}^d.
$$

Thus to see the reason why the last column of $\mathbf{V}$ has $\mathbf{A}^d_{1k}, \mathbf{A}^d_{2k}, \ldots$ as its first k rows (these are the coefficients that multiply with $X_n, X_{n+1}, \ldots, X_{n+k-1}$) we only need to see that from the above matrix equation, $X_{n+d+k-1}$ is the dot product,

$$
X_{n+d+k-1} = \begin{bmatrix} X_n & X_{n+1} & \ldots & X_{n+k-1} \end{bmatrix} \cdot \mathbf{A}^d_{*k},
$$

where $\mathbf{A}^d_{*k}$ denotes the kth column of $\mathbf{A}^d$.

The above outlines how the matrix $\mathbf{V}$ arises, but as the computations show, and the expressions $\mathbf{A}_{ik}^j$ show we need to compute the powers of the matrix $\mathbf{A}$. Computing the powers of the matrix is an expensive computation. The new, more efficient method we explain later avoids this and instead comes up with a basis that is immediate and requires no computation!

We now proceed with discussing the algorithm for the spectral distance. The corresponding dual basis matrix for $\mathbf{V}$ is

$$\mathbf{W} = \begin{bmatrix} p\mathbf{I}_{k \times k} & \mathbf{0}_{k \times d} \\ -\mathbf{X}'_{d \times k} & \mathbf{I}_{d \times d} \end{bmatrix}, \tag{7.1}$$

which is orthogonal to basis matrix $\mathbf{V}$ because $\mathbf{VW}' = p\mathbf{I}$. Notice that the determinant of $\mathbf{W}$ is p^k. Notice that any set of $k + d$ linearly independent vectors from the dual lattice with the corresponding determinant equal to p^k form a basis for the dual lattice.

We remark that, to obtain the elements $[\mathbf{A}^i]_{jk}$ for $i = 2, 3, \ldots, d, j = 1, 2, \ldots, k$, it is not necessary to compute the $k \times k$ matrix $\mathbf{A}^i, i = 2, 3, \ldots, d$. It was shown in L'Ecuyer and Couture (1997) that, for each $j = 1, 2, \ldots, k$, $\{[\mathbf{A}^i]_{jk}, i = 1, 2, \ldots, d\}$ are the first d numbers generated from the MRG using the k-dimensional unit vector $e_{i(k)}$ (i.e., the k-dimensional vector with 1 in the i-th coordinate and 0 elsewhere) as the seed. Note that $[\mathbf{A}]_{jk} = \alpha_{k-j+1}$ for $j = 1, 2, \ldots, k$.

The length of the shortest vector in the dual space can be found by solving the following minimization problem:

$$v_{k+d}^2(k) = \min_{(c_1, c_2, \ldots, c_d) \neq (0, 0, \ldots, 0)} \left(\left\| \left[\sum_{i=1}^d c_i \, \mathbf{w}_i \right]_p \right\|^2 \right), \tag{7.2}$$

where $c_i \in \{0, \ldots, (p-1)\}$ and

$$\mathbf{w}_1 = (-\alpha_k, -\alpha_{k-1}, \ldots, -\alpha_1, 1, 0, 0, \ldots, 0)',$$
$$\mathbf{w}_2 = (-[\mathbf{A}^2]_{1k}, -[\mathbf{A}^2]_{2k}, \ldots, -[\mathbf{A}^2]_{kk}, 0, 1, \ldots, 0)' \pmod{p},$$
$$\vdots$$
$$\mathbf{w}_d = (-[\mathbf{A}^d]_{1k}, -[\mathbf{A}^d]_{2k}, \ldots, -[\mathbf{A}^d]_{kk}, 0, 0, \ldots, 1)' \pmod{p}.$$

The minimization problem in Eq. 7.2 is usually solved by submitting $\mathbf{W}$ to some basis reduction algorithm. The spectral distance $d_t(k)$ is equal to the reciprocal of $v_{k+d}(k)$, the length of the shortest vector in the reduced basis. We remark that the basis matrix $\mathbf{V}$ is not directly related to the computation of the spectral test. Rather, it is only used for finding the dual matrix $\mathbf{W}$.

Notice that this method always passes a $(k + d) \times (k + d)$ matrix to LLL (or to another lattice basis reduction algorithm). Although the method in L'Ecuyer and Couture (1997) as described above is straightforward to obtain the dual matrix $\mathbf{W}$, it may not be efficient when k is large because it requires to generate k different subsequences of length d from Eq. 6.1. Furthermore, a $(k + d) \times (k + d)$ dual basis matrix $\mathbf{W}$ must be submitted to a lattice basis reduction algorithm. It is widely known that the Fincke-Pohst algorithm is exponential in time and the LLL algorithm is polynomial in time; (see, e.g., Cohen 1993). When k is large, the lattice basis reduction of $\mathbf{W}$ can be extremely costly in both time and computer resources regardless of which lattice basis algorithm is used. In the following subsection, we propose a novel method that provides a fast way to create an equivalent yet smaller dual basis matrix—and therefore, more computationally efficient—for certain classes of MRGs.

7.2   Efficient Method for MRG's Spectral Test

As discussed in Chap. 6, to compute the spectral test of an MRG, we consider all the successive sequences of length t generated from the recurrence equation: $\mathbf{S}_n = (X_n, X_{n+1}, \ldots, X_{n+t-1})'$ for $n = 0, 1, \ldots, \rho - 1$, where $\rho = p^k - 1$ is the period length. $\mathbf{S}_n$ is referred to as the state of the MRG at step n. Let I be a set of fixed nonnegative integers. The particular set of integers $I = \{0, 1, \ldots, t - 1\}$ could be thought of as the indices of the state $\mathbf{S}_n$, which create all the possible successive t-tuples over all steps, $n = 0, 1, \ldots, \rho - 1$, in the entire period of the MRG.

The index set I is not required to be a set of successive integers; more generally, it can be any set of nonnegative integers $I = \{j_1, j_2, \ldots, j_r\}$, where $j_1 < j_2 < \cdots < j_r$ and r is the number of indices in I. Let $t = j_r - j_1 + 1$, the number of variates that need to be generated at each step of the MRG state $\mathbf{S}_n$. Denote the set of all possible r-tuples that a maximum-period MRG can generate according to I by

$$L_r(I) = \left\{ (X_{n+j_1}, X_{n+j_2}, \ldots, X_{n+j_r}) \mid n = 0, 1, \ldots, \rho - 1 \right\}. \tag{7.3}$$

7.2.1   Spectral Test When $t = k + 1$

We can naturally extend the algorithm for computing the spectral test value of an LCG described earlier to an MRG of order k as defined in Eq. 6.1 for dimension $t = k + 1$. First, consider $L_{k+1}(I)$ as in Eq. 7.3 for $I = \{0, 1, \ldots, k\}$, in which the $(k + 1)$-tuples are generated from an MRG of order k. Next, note that the points in $\Lambda_{k+1}(I)$ form a $(k + 1)$-dimensional lattice over $[0, 1)^{k+1}$ that can be covered by a family of parallel k-dimensional hyperplanes with the normal vector $\mathbf{w} = (\alpha_k, \alpha_{k-1}, \ldots, \alpha_1, -1)'$.

Clearly, for each integer c, every $(k + 1)$-tuple in $L_{k+1}(I)$ also satisfies the equation

$$cX_i = ([c\alpha_1]_p X_{i-1} + \cdots + [c\alpha_k]_p X_{i-k}) \bmod p, \quad i \geq k.$$

Hence, each choice of c defines a (not necessarily unique) family of parallel k-dimensional hyperplanes that cover all the points in $\Lambda_{k+1}(I)$ with the normal vector being $[c\,\mathbf{w}]_p$.

Therefore, among all choices of c, we can find the shortest normal vector by computing

$$v_{k+1}^2(k) = \min_{c \neq 0} \left\| [c \, \mathbf{w}]_p \right\|^2 = \min_{0 < c < p} \left(\sum_{i=1}^{k} [c\alpha_i]_p^2 + c^2 \right). \tag{7.4}$$

Note that the symmetry property allows us to cut down the search space for the integer c by half to $0 < c < p/2$. A similar result was given in L'Ecuyer and Simard (2014), in which the authors applied the result to MRGs with few nonzero terms. Clearly, the spectral test value $v_{k+1}^2(k)$ depends *only* on the set of nonzero α_i values. Once $v_{k+1}^2(k)$ is computed, the spectral distance for dimension $k+1$ is given by

$$d_{k+1}(k) = \frac{1}{v_{k+1}(k)}.$$

For any dimension t, a larger $v_t(k)$ corresponds to a smaller $d_t(k)$, which implies more uniform coverage of the t-tuples. Therefore, for finding a good MRG of a given order k, it is desirable to have $v_t(k)$ as large as possible. Since the primary concern in computing the spectral test is to find the normal vector with the shortest squared length $v_t^2(k)$, we can designate $v_t^2(k)$ as the value representing the spectral test, for which the larger the better. We will simply call $v_t^2(k)$ the *spectral test value* hereafter.

The Algorithm 1 presents the steps to compute the spectral test value $v_{k+1}^2(k)$, for an MRG of order k. It is simple and straightforward but only works for the case of $t = k + 1$. In Step 2.c., we can stop when $v_{\min}^2 \leq (c+1)^2$ because $v_j^2 = j^2 + \sum_{i=1}^{k} [c\alpha_i]_p^2 \geq j^2$ for all j which implies $v_j^2 \geq v_{\min}^2$ for all $j \geq (c+1)$. For t greater than $k+1$, as we will see next, we would need a more sophisticated algorithm.

Algorithm 1: Algorithm to compute the spectral test value $v_{k+1}^2(k)$ for an MRG of order k

1. Initially, for $c = 1$, set $v_{\min}^2 = 1 + \sum_{i=1}^{k} \alpha_i^2$.
2. For $c = 2, 3, \ldots,$ do

   a. compute $v_c^2 = c^2 + \sum_{i=1}^{k} [c\alpha_i]_p^2$;
   b. if $v_{\min}^2 > v_c^2$, then reset $v_{\min}^2 = v_c^2$;
   c. if $v_{\min}^2 \leq (c+1)^2$, then break; else continue with the next c;

3. Deliver $v_{k+1}^2(k) = v_{\min}^2$.

7.2.2   Spectral Test When $t > k + 1$

As presented in the previous subsection, the method of computing the spectral test value for $t = k + 1$ is to find the normal vector with the shortest length via solving a minimization problem. We can extend this method to dimension $t > k + 1$ as follows.

For dimension $t = k + d$ with integer $d > 1$, we first define the following d normal vectors of length t:

$$
\begin{aligned}
\mathbf{w}_1 &= (\alpha_k, \alpha_{k-1}, \ldots, \alpha_1, -1, 0, 0, \ldots, 0)', \\
\mathbf{w}_2 &= (0, \alpha_k, \alpha_{k-1}, \ldots, \alpha_1, -1, 0, \ldots, 0)', \\
&\ \ \vdots \\
\mathbf{w}_d &= (0, 0, \ldots, 0, \alpha_k, \alpha_{k-1}, \ldots, \alpha_1, -1)'.
\end{aligned}
\tag{7.5}
$$

Note that $\mathbf{w}_{i+1}$ is merely a simple rotation of $\mathbf{w}_i$ for $i = 1, 2, \ldots, d - 1$. In addition, the d normal vectors are linearly independent of each other, and they all are orthogonal to $\mathbf{S}_n = (X_n, X_{n+1}, \ldots, X_{n+t-1})'$, the state of the MRG at step n, for any $n \geq 0$, where $t = k + d$.

More specifically, $\mathbf{S}_n' \mathbf{w}_i = \alpha_k X_{n+i-1} + \alpha_{k-1} X_{n+i} + \cdots + \alpha_1 X_{n+k+i-2} - X_{n+k+i-1} = X_{n+k+i-1} - X_{n+k+i-1} = 0$ for any $n \geq 0$ and $i = 1, 2, \ldots, d$. We then solve the following minimization problem:

$$
v_{k+d}^2(k) = \min_{(c_1, c_2, \ldots, c_d) \neq (0, 0, \ldots, 0)} \left(\left\| \left[\sum_{i=1}^{d} c_i \, \mathbf{w}_i \right] \right\|_p^2 \right),
\tag{7.6}
$$

$0 \leq c_1, c_2, \ldots, c_d < p$. The minimum value $v_{k+d}^2(k)$ depends on the choices of the MRG's multipliers $\alpha_1, \alpha_2, \ldots, \alpha_k$. Clearly, $v_{k+1}^2(k) \geq v_{k+2}^2(k) \geq \cdots \geq v_{k+d}^2(k)$; that is, as d gets larger, the uniform spread of the t-tuples can at best stay the same, if not getting worse. However, solving this minimization problem becomes increasingly difficult for MRGs as d increases. Even more difficult is searching for the "best" multipliers (α_i's) such that $v_t^2(k)$ is the largest for the given order k and modulus p.

In summary, computing the spectral test for MRGs is equivalent to solving the aforementioned minimization problem of finding the shortest normal vector. In the next section, we will develop a new method for computing the spectral test. As we will explain, this method is easier to implement than the method that is commonly used.

7.2.3   Spectral Test When $t > k$

From a basis of a lattice, one can find another basis whose vectors are relatively shorter and nearly "orthogonal." This process is called lattice basis reduction (see, e.g., Cohen

1993). We will use lattice basis reduction to help compute the t-dimensional spectral test, where $t = k + d$ with integer $d > 0$.

Since $L_t(I)$ in Eq. 7.3 with $I = \{0, 1, \ldots, t - 1\}$ by itself is not a lattice, it is common to "extend" it in all "directions" by adding every vector in $L_t(I)$ to every vector in $p\mathbb{Z}^t$, where $\mathbb{Z}^t$ denotes the set of t-dimensional vectors of integers. Hence, from $L_t(I)$ one can create a lattice, namely, $L_t(I) + p\mathbb{Z}^t$, which is the periodic continuation of $L_t(I)$ with period p. Define a $t \times t$ matrix $\mathbf{V}$ whose rows are basis vectors of $L_t(I) + p\mathbb{Z}^t$. A dual space to $L_t(I) + p\mathbb{Z}^t$ exists with a dual $t \times t$ basis matrix $\mathbf{W}$ such that $\mathbf{VW}' = p\mathbf{I}$, where $\mathbf{I}$ is the $t \times t$ identity matrix. The length of the shortest vector in this dual space is equal to $v_t(k)$ (see, e.g., L'Ecuyer and Couture 1997). Therefore, we can find $v_t(k)$ by performing lattice basis reduction on $\mathbf{W}$.

To perform this lattice basis reduction, we can use the help of the popular LLL algorithm proposed by Lenstra et al. (1982). Essentially, the LLL algorithm is an integer lattice version of Gram-Schmidt orthogonalization. Several LLL implementations are available in software packages such as MAPLE, NTL, and Sage (an open-source computer algebra system). Although the LLL Algorithm does not always (but often does) yield the shortest vector, we can still use it to obtain an approximation of the spectral test, which allows us to find "better" pseudo-random number generators; see Entacher et al. (2002) for more details. We will use LLL for this purpose in Sect. 7.4.

To guarantee the shortest vector to be found, the exact enumeration algorithm of Fincke-Pohst algorithm (Fincke and Pohst 1985) can be used; (see, e.g., L'Ecuyer and Couture 1997, L'Ecuyer et al. 1993). For discussions on differences between the LLL Algorithm and Fincke-Pohst algorithm, (see, e.g., Entacher et al. 2002) and the references cited therein. The method for finding $v_{k+d}^2(k)$ given in the next subsection can be coupled with either algorithm. In Sect. 7.2.4, we will discuss how the proposed method for creating $\mathbf{W}$ differs from the commonly used method given in the literature.

Denote the simple rotation operation as shown in Eq. 7.5 by $\mathbf{R}(\cdot)$.

Together, Steps 1 and 2 create our initial $d \times t$ matrix of normal vectors, which simply are rotations of $\mathbf{w}_1$. Columns of zeros in $\mathbf{M}_0$ are removed in Step 3, if there are any, because they play no role in solving the minimization problem. If some columns of zeros are removed, then it is at this step that we essentially "reduce the size" of the matrix to be submitted to a lattice basis reduction algorithm. Notice that, this idea of dropping columns of zeros, is potentially applicable to the existing method described in Sect. 7.1. However, the dual basis matrix constructed by the algorithm (see Eq. 7.1) does not allow such reduction, because no column is entirely zeros. We will give an illustrative example below. Then later in Sect. 7.3.2, we will give an example where there is a dramatic decrease in matrix size—as we will see, had there not been a dramatic decrease in matrix size, the lattice basis reduction would have been very difficult to complete for such a large t. Step 4 adds the necessary number of unit vectors (times p) such that the rows of $\mathbf{M}$ form a basis for the dual space spanned by the normal vectors. Finally, Step 5 uses any preferred algorithm to find the value or an upper bound of the spectral test value $v_t^2(k)$.

Algorithm 2: Algorithm for computing/approximating the value of $v_{k+d}^2(k)$

---

1. **Create the initial d normal vectors.** Let $\mathbf{w}_1 = (\alpha_k, \alpha_{k-1}, \ldots, \alpha_1, -1, 0, \ldots, 0)'$, in which the last $d - 1$ entries are all zero. Compute the remaining $d - 1$ normal vectors by $\mathbf{w}_i = \mathbf{R}^{i-1}(\mathbf{w}_1)$ for $i = 2, 3, \ldots, d$.

2. **Create the initial matrix.** Let $\mathbf{M}_0$ be the initial $d \times t$ matrix whose d rows are $\mathbf{w}_1', \mathbf{w}_2', \ldots, \mathbf{w}_d'$.

3. **Remove columns of zeros.** Remove any columns of all zeros from the initial matrix $\mathbf{M}_0$. Let $\mathbf{M}_1$ be the $d \times t^*$ matrix formed by the remaining t^* columns. If there are no columns of zeros in $\mathbf{M}_0$, then $t^* = t$ and $\mathbf{M}_1 = \mathbf{M}_0$.

4. **Create the final matrix for basis reduction.** Let $\mathbf{M}$ be the $t^* \times t^*$ matrix with the i-th row of the first $t^* - d$ rows being $p\mathbf{e}'_{i(t^*)}$ for $i = 1, 2, \ldots, t^* - d$, where p is the modulus and $\mathbf{e}_{i(t^*)}$ is the i-th unit vector of dimension t^*. Fill the remaining d rows with the rows of $\mathbf{M}_1$.

5. **Compute/approximate the value of $v_{k+d}^2(k)$.** Apply a basis reduction algorithm to matrix $\mathbf{M}$ to yield a reduced matrix $\mathbf{M}^*$ consisting of a set of "reduced" *dual* basis vectors. Then compute the squared length of the shortest row vector in $\mathbf{M}^*$, which is related to the spectral test value $v_{k+d}^2(k)$. As mentioned earlier, when the popular LLL algorithm is applied, the shortest row vector in $\mathbf{M}^*$ obtained may not always be the shortest vector in the space spanned by the row vectors of $\mathbf{M}^*$ (which in turn correspond to the shortest vector in the lattice), although it often is. To guarantee the shortest vector in $\mathbf{M}^*$ indeed corresponds to the shortest vector in the lattice, we can apply an exact (but less efficient) enumeration algorithm such as the branch-and-bound algorithm of Fincke and Pohst (1985) for basis reduction.

We now outline the proof of correctness for this algorithm. First, we notice that ignoring the columns of zeros in step 3 does not affect the correctness of the algorithm since eventually, we need to minimize a certain linear combination of vectors after a modulo p operation, and the column of zeros represents a coordinate that is zero on all the vectors. As such we assume for simplicity that this is not done. This allows us to talk in the initial space of $t = k + d$ dimensions. It is not hard to see that the vectors $p\mathbf{e}_{1(t)}, p\mathbf{e}_{2(t)}, \ldots, p\mathbf{e}_{k(t)}$ along with $\mathbf{w}_1, \ldots, \mathbf{w}_d$ are all linearly independent. Moreover, it can also be verified in a straightforward manner that they all belong to the dual lattice. But, this is not enough to prove that they are a basis for the dual lattice. However, since the determinant of the matrix $\mathbf{M}_0$ is p^k, as can be easily verified, it is equal to the determinant of the dual basis as shown in Sect. 7.1. This shows that $\{p\mathbf{e}_{1(t)}, p\mathbf{e}_{2(t)}, \ldots, p\mathbf{e}_{k(t)}, \mathbf{w}_1, \ldots, \mathbf{w}_d\}$ form a basis of the dual lattice. Moreover, similar to the case of the vectors $p\mathbf{e}_{1(t)}, p\mathbf{e}_{2(t)}, \ldots, p\mathbf{e}_{k(t)}$, the vectors $p\mathbf{e}_{(k+1)(t)}, \ldots, p\mathbf{e}_{t(t)}$ also belong to the dual lattice. Therefore, they can be used to form linear combinations of the basis vectors. For fixed values of $c_1, \ldots, c_d$, let $\mathbf{w} = \sum_{j=1}^d c_j \mathbf{w}_j$. Consider an arbitrary linear combination, $\lambda_1 \mathbf{e}_{1(t)} + \cdots + \lambda_k \mathbf{e}_{k(t)} + c_1 \mathbf{w}_1 + \cdots + c_d \mathbf{w}_d = \lambda_1 \mathbf{e}_{1(t)} + \cdots + \lambda_k \mathbf{e}_{k(t)} + \mathbf{w}$, of the dual lattice basis vectors, where the $\lambda_i, c_j \in \mathbb{Z}$. For a coordinate position $\ell \in \{1, \ldots, t\}$, let $\mathbf{w}^\ell$ denote ℓth coordinate of $\mathbf{w}$. Then, the ℓth coordinate of $\lambda_1 \mathbf{e}_{1(t)} + \cdots + \lambda_k \mathbf{e}_{k(t)} + \mathbf{w}$ is in the same congruence class modulo p as

$\mathbf{w}^\ell$ since the other vectors, namely, $p\mathbf{e}_{1(t)}, p\mathbf{e}_{2(t)}, \ldots, p\mathbf{e}_{k(t)}$ either have a 0 or p in each coordinate position. Thus, the minimum possible absolute value that can be achieved for coordinate ℓ in $\lambda_1\mathbf{e}_{1(t)} + \cdots + \lambda_k\mathbf{e}_{k(t)} + \mathbf{w}$ is at least $\left[\mathbf{w}^\ell\right]_p$. This can be easily done for $\ell \in \{1, \ldots, k\}$ since the vectors $p\mathbf{e}_{1(t)}, p\mathbf{e}_{2(t)}, \ldots, p\mathbf{e}_{k(t)}$ can be used to achieve any number in the congruence class of $\mathbf{w}^\ell$ modulo p. But, indeed this can also be done for $\ell \in \{k+1, \ldots, t\}$ since we may assume that the dual lattice vectors $p\mathbf{e}_{(k+1)(t)}, \ldots, p\mathbf{e}_{t(t)}$ are also used in linear combinations, along with the basis vectors. Achieving the minimum absolute value for all coordinate positions $\ell \in \{1, \ldots, t\}$ we will get the vector $[\mathbf{w}]_p$, and this is also the vector with the minimum norm in the dual lattice for fixed $c_1, \ldots, c_d$. Thus to find the minimum norm nonzero vector in the dual lattice we need to solve the minimization problem,

$$v_{k+d}^2(k) = \min_{(c_1, c_2, \ldots, c_d) \neq (0, 0, \ldots, 0)} \left(\left\| \left[\sum_{i=1}^d c_i \, \mathbf{w}_i \right]_p \right\|^2 \right).$$

Clearly, we may assume that each c_i is in the range $0 \le c_i < p$, as we can always adjust arbitrary c_i to lie in this range without changing the value obtained for any coordinate position modulo p.

The minimization is done by passing the matrix $\mathbf{M}$ to a lattice basis reduction algorithm.

Example 7.1 (*A simple illustrative example*)

To illustrate Algorithm 2, consider the case of $d = 3$ for MRGs of order 5:

$$X_i = (\alpha_1 X_{i-1} + \alpha_2 X_{i-2} + \alpha_3 X_{i-3} + \alpha_4 X_{i-4} + \alpha_5 X_{i-5}) \bmod p, \quad i \ge 5.$$

The first of the d normal vectors is

$$\mathbf{w}_1 = (\alpha_5, \alpha_4, \alpha_3, \alpha_2, \alpha_1, -1, 0, 0)'.$$

The normal vectors $\mathbf{w}_2$ and $\mathbf{w}_3$ can be easily obtained by rotating the elements in $\mathbf{w}_1$ once and twice, respectively. Then we have

$$\mathbf{M}_0 = \begin{bmatrix} \alpha_5 & \alpha_4 & \alpha_3 & \alpha_2 & \alpha_1 & -1 & 0 & 0 \\ 0 & \alpha_5 & \alpha_4 & \alpha_3 & \alpha_2 & \alpha_1 & -1 & 0 \\ 0 & 0 & \alpha_5 & \alpha_4 & \alpha_3 & \alpha_2 & \alpha_1 & -1 \end{bmatrix}.$$

For a two-term MRG with $\alpha_2 = \alpha_3 = \alpha_4 = 0$, the 4th column in $\mathbf{M}_0$ is a zero column and can be removed. If $\mathbf{M}_0$ has no columns of all zeros, we have $\mathbf{M}_1 = \mathbf{M}_0$ and the final matrix to be submitted to LLL or any basis reduction algorithm is

$$M = \begin{bmatrix} p & 0 & 0 & 0 & 0 & 0 & 0 & 0 \\ 0 & p & 0 & 0 & 0 & 0 & 0 & 0 \\ 0 & 0 & p & 0 & 0 & 0 & 0 & 0 \\ 0 & 0 & 0 & p & 0 & 0 & 0 & 0 \\ 0 & 0 & 0 & 0 & p & 0 & 0 & 0 \\ \alpha_5 & \alpha_4 & \alpha_3 & \alpha_2 & \alpha_1 & -1 & 0 & 0 \\ 0 & \alpha_5 & \alpha_4 & \alpha_3 & \alpha_2 & \alpha_1 & -1 & 0 \\ 0 & 0 & \alpha_5 & \alpha_4 & \alpha_3 & \alpha_2 & \alpha_1 & -1 \end{bmatrix}.$$

This example showcases the simplicity of the proposed method in finding a dual basis matrix (i.e., the matrix M). Both the "traditional" W matrix and the proposed M matrix are bases for the same dual space. For large-order MRGs (with many nonzero terms), a large matrix composed of dual basis vectors would need to be created. The aforementioned algorithm can create this matrix rather efficiently.

7.2.4   Comparison with the Existing Method

When $t = k + 1$, our proposed method solves a similar minimization problem to that of the existing method. However, for $t > k + 1$, the existing method may require submitting a large $(k + d) \times (k + d)$ matrix to a lattice basis reduction algorithm. When k is large, as mentioned earlier, this may not be feasible due to the time complexity of lattice basis algorithms. However, for large-order MRGs with few nonzero terms, our method can find a much smaller matrix that yields the same spectral test value. Furthermore, the proposed method is more intuitive than the existing method in explaining what is being minimized and how the minimization problem relates to the spectral test. Specifically, we are searching among all the families of parallel hyperplanes that cover all the t-tuples as in Eq. 7.3 for a family with the shortest normal vector; and the spectral distance $d_t(k)$ is the reciprocal of the length of this shortest normal vector $v_t(k)$.

Overall, our approach is similar to that of the existing method in the way that both methods find a dual basis matrix and submit it to a basis reduction algorithm. Nonetheless, we propose a simpler way to find a set of dual basis to the lattice points created from $L_t(I)$ in Eq. 7.3.

Lastly, as we will see in the next section, for certain special classes of large-order MRGs with few nonzero terms, the proposed method can be substantially more efficient than the existing method.

7.3   Spectral Test for DX-K-S Generators

Deng and Xu (2003) proposed a class of portable, efficient, and maximum-period MRGs called DX-k-s generators as follows:

1. when $\alpha_1 = 1, \alpha_k = B$, it is a DX-$k$-1 or an FMRG (Fast MRG) as considered in Deng and Lin (2000):

$$X_i = X_{i-1} + BX_{i-k} \bmod p, \quad i \geq k;$$

2. when $\alpha_1 = B, \alpha_k = B$, it is a DX-$k$-2 generator with two nonzero terms of the same multiplier:

$$X_i = B(X_{i-1} + X_{i-k}) \bmod p, \quad i \geq k;$$

3. when $\alpha_1 = \alpha_{\lceil k/2 \rceil} = \alpha_k = B$, it is a DX-$k$-3 generator with three nonzero terms:

$$X_i = B(X_{i-1} + X_{i-\lceil k/2 \rceil} + X_{i-k}) \bmod p, \quad i \geq k;$$

4. when $\alpha_1 = \alpha_{\lceil k/3 \rceil} = \alpha_{\lceil 2k/3 \rceil} = \alpha_k = B$, it is a DX-$k$-4 generator with four nonzero terms:

$$X_i = B(X_{i-1} + X_{i-\lceil k/3 \rceil} + X_{i-\lceil 2k/3 \rceil} + X_{i-k}) \bmod p, \ i \geq k.$$

Here the notation $\lceil x \rceil$ is the ceiling function of a number x, returning the smallest integer $\geq x$. For DX-k-s generators, s is the number of terms sharing the multiplier B.

7.3.1   Spectral Test for DX-k-s When $t = k + 1$

When computing the spectral test value $v_{k+1}^2(k)$ for FMRG-k generators, the normal vector $\mathbf{w} = (B, 0, \ldots, 0, 1, -1)'$, consisting of one B, one 1, one -1, and $(k - 2)$ 0's. For DX-k-s generators, the normal vector $\mathbf{w}$ consists of s B's, one -1, and $(k - s)$ 0's. Then the minimization problem in Eq. 7.4 can be further simplified to

$$v_{k+1}^2(k) = \min_{0<c<p/2} \left([cB]_p^2 + 2c^2\right) \tag{7.7}$$

for FMRG-k generators, and

$$v_{k+1}^2(k) = \min_{0<c<p/2} \left(s\,[cB]_p^2 + c^2\right) \tag{7.8}$$

for DX-k-s generators. For convenience, we will mainly discuss the minimization problem for DX-k-s generators in Eq. 7.8. Nevertheless, the results would be similar for FMRG-k

generators. We remark that equivalent expressions to Eqs. 7.7 and 7.8 were also given in L'Ecuyer and Simard (2014).

It is interesting to note that $v^2_{k+1}(k)$ in Eq. 7.8 depends only on the parameters B and s, but not on the order k, a result of great importance for the efficiency of our method.

Also note that for the LCG, the expression for $v^2_{k+1}(k)$ as in Eq. 7.8 is simplified to

$$v^2_2 = \min_{0<c<p/2} \left([cB]^2_p + c^2 \right). \qquad (7.9)$$

The lattice structure for the LCG has been studied extensively in the literature (see, e.g., Entacher et al. 2005, Sezgin 1996, 2004, 2006). From these studies, it is known that the LCG with multiplier B tends to have a bad (i.e., small) v^2_2 whenever B is close to the value of pN/D with small numerator N and/or small denominator D. One reason is that, under such circumstances, by taking $c = D$, $c \times B \mod p$ will yield a small value (see Sezgin 2004, for more details). Based on the similarity between Eqs. 7.7, 7.8, and 7.9, the same argument of having a bad lattice structure for LCGs can also be applied to FMRG-k and DX-k-s generators. This finding helps us avoid choosing certain poorly-performed values for B.

We remark that, for higher values of $t = k + d$, the lattice structure of an LCG would get worse, whereas that of FMRG-k and DX-k-s generators would have a high chance to remain the same, as to be shown in Sect. 7.3.3. Before that, let us show how to compute the spectral test value when $t = k + 2$ for DX-k-s generators; the method for FMRG-k generators is similar and hence omitted.

7.3.2 Spectral Test for DX-k-s When $t = k + 2$

For DX-k-s generators, the minimization problem in Eq. 7.6 for $t = k + 2$ can be simplified to

$$v^2_{k+2}(k) = \min_{(c_1,c_2)\neq(0,0)} \left(s\,[c_1B]^2_p + (s-1)\,[c_2B]^2_p + c^2_2 + [-c_1+c_2B]^2_p \right), \qquad (7.10)$$

where the possible range of c_1 and c_2 can be set as $-p/2 < c_1, c_2 < p/2$. Similar to the case for finding $v^2_{k+1}(k)$, it is pretty straightforward to solve this minimization problem. However, we can also compute $v^2_{k+2}(k)$ using the algorithm given in Sect. 7.2.1.

As an illustrative example, consider a DX-25013-2 generator. For $t = k + 2$, the two 25015-dimensional normal vectors are

$$\mathbf{w}_1 = (B, 0, 0, \ldots, 0, B, -1, 0)',$$
$$\mathbf{w}_2 = (0, B, 0, 0, \ldots, 0, B, -1)'.$$

By removing the $25010 (= k - 3)$ columns of zeros in $\mathbf{M}_0 = [\ \mathbf{w}_1,\ \mathbf{w}_2]'$, we have

$$\mathbf{M}_1 = \begin{bmatrix} B & 0 & B & -1 & 0 \\ 0 & B & 0 & B & -1 \end{bmatrix}.$$

Note that the number of columns has been drastically reduced from $25015 (= t = k + 2)$ of the initial matrix $\mathbf{M}_0$ down to 5. Next, to compute $v_{k+2}^2(k)$, we simply apply basis reduction algorithm to the much smaller matrix

$$\mathbf{M} = \begin{bmatrix} p & 0 & 0 & 0 & 0 \\ 0 & p & 0 & 0 & 0 \\ 0 & 0 & p & 0 & 0 \\ B & 0 & B & -1 & 0 \\ 0 & B & 0 & B & -1 \end{bmatrix}.$$

This example demonstrates that our proposed method is able to decrease the size of a prohibitively large 25015×25015 dual basis matrix to an equivalent 5×5 matrix. Without this dramatic decrease in size, the lattice basis reduction would have been very difficult to complete.

Note that the special structure of DX-k-s generators allows us to compute the spectral test in a very efficient way no matter how large the order k is. As we will show next, this special structure also gives the consistency property described earlier.

7.3.3   Consistency Property of DX-k-s

As noted earlier, $v_{k+1}^2(k) \geq v_{k+2}^2(k)$ for any MRG of order k. Interestingly, for many (but not all) DX-k-s generators, we have observed that

$$v_{k+1}^2(k) = v_{k+2}^2(k).$$

We will call this property the *consistency* of the spectral test values. As a simple experiment, we randomly selected 5000 B's as potential multiplier of the DX-k-s generator. It turns out that $82.92, 96.24$, and 100% of the tested DX-k-s generators have the consistency property for $s = 2, 3, 4$, respectively. Since there are exceptions, it seems hard to have a general theory for the consistency. Nonetheless, the following are some intuitive explanations for this phenomenon, but are not meant to be very precise.

1. Our first explanation is algebraic. Notice that by Eq. 7.8 we have that,

$$v_{k+1}^2(k) = \min_{0 < c < p/2} \left(s\, [cB]_p^2 + c^2 \right) = \min_{0 < c < p/2} \left\| [c\,\mathbf{w}_1]_p \right\|^2,$$

and by Eq. 7.10 we have that,

$$v_{k+2}^2(k) = \min_{(c_1,c_2)\neq(0,0)} \left(s\,[c_1 B]_p^2 + (s-1)\,[c_2 B]_p^2 + c_2^2 + [-c_1 + c_2 B]_p^2 \right)$$

$$= \min_{(c_1,c_2)\neq(0,0)} \left\| [c_1\,\mathbf{w}_1 + c_2\,\mathbf{w}_2]_p \right\|^2 .$$

Now, in general,

$$[c_1\,\mathbf{w}_1 + c_2\,\mathbf{w}_2]_p \neq [c_1\,\mathbf{w}_1]_p + [c_2\,\mathbf{w}_2]_p ,$$

however for the vectors $[c_1\,\mathbf{w}_1]_p$, $[c_2\,\mathbf{w}_2]_p$, for all but one coordinate position, at least one of the vectors has a 0 coordinate. In only 1 coordinate position one of the vectors is $[-c_1 + c_2 B]_p$ while the other has, $-c_1 + [c_2 B]_p$. If $[-c_1 + c_2 B]_p$ is close to $-c_1$ then $v_{k+2}^2(k)$ is expected to be close to $s\,[c_1 B]_p^2 + c_1^2$ and since $v_{k+1}^2(k)$ is the minimum value of this expression over all c_1, $v_{k+2}^2(k)$ is expected to be close to $v_{k+1}^2(k)$. Notice that here we have also the terms $(s-1)\,[c_2 B]_p^2 + c_2^2$ acting to increase the value of $v_{k+2}^2(k)$. Moreover, the effect of this increase is more as s increases. The same argument goes if $[-c_1 + c_2 B]_p$ is close to $[c_2 B]_p$. For $v_{k+2}^2(k)$ to be much smaller than $v_{k+1}^2(k)$ we expect that $[-c_1 + c_2 B]_p$ is much smaller in magnitude than $[-c_1]_p$ or $[c_2 B]_p$. For a fixed B, we expect this to be a "rare" event (over random choices of c_1, c_2), and in particular this may not be met for the minimizing choices of c_1, c_2. This is consistent with the empirical finding mentioned earlier that the percentage of DX-k-s with $v_{k+2}^2(k) < v_{k+1}^2(k)$ decreases from 17.08% for $s = 2$ to 3.76% for $s = 3$ and further down to 0% for $s = 4$. Of course, the actual percentages would vary with another sample of the multiplier B.

2. Our second intuitive reason is geometric. We may consider that

$$\left\| [c_1\,\mathbf{w}_1 + c_2\,\mathbf{w}_2]_p \right\| \approx \left\| [c_1\,\mathbf{w}_1]_p + [c_2\,\mathbf{w}_2]_p \right\| ,$$

since those vectors differ in at most 1 coordinate as remarked above. Indeed, one can show easily that,

$$\left\| [c_1\,\mathbf{w}_1 + c_2\,\mathbf{w}_2]_p \right\|^2 \leq \left\| [c_1\,\mathbf{w}_1]_p \right\|^2 + \left\| [c_2\,\mathbf{w}_2]_p \right\|^2 \leq \left\| [c_1\,\mathbf{w}_1 + c_2\,\mathbf{w}_2]_p \right\|^2 + (p^2 - 2p).$$

Thus in what follows we instead think of the minimization as,

$$v_{k+2}^2(k) \approx \min_{(c_1,c_2)\neq(0,0)} \left\| [c_1\,\mathbf{w}_1]_p + [c_2\,\mathbf{w}_2]_p \right\|^2 .$$

The reason for this viewing is that it is of course easier to discuss simple addition of the vectors $[c_1\,\mathbf{w}_1]_p$, $[c_2\,\mathbf{w}_2]_p$ as opposed to the non-linear operation of $[c_1\,\mathbf{w}_1 + c_2\,\mathbf{w}_2]_p$. Now, for DX-$k$-$s$ generators, the vectors $[c_1\,\mathbf{w}_1]_p$ and $[c_2\,\mathbf{w}_2]_p$ are "nearly" orthogonal (perpendicular) to each other, because there is only one coordinate position that both are nonzero in and the inner product of these two vectors, $[c_1\,\mathbf{w}_1]_p$ and $[c_2\,\mathbf{w}_2]_p$, is $-c_1\,[c_2 B]_p$. Now, geometrically, for any two vectors, $\mathbf{y}$ and $\mathbf{z}$, $\|\mathbf{y} + \mathbf{z}\|^2$ can be smaller than $\|\mathbf{y}\|^2$ or $\|\mathbf{z}\|^2$ only when the "angle" between the two vectors is much larger than 90°C. Since the two vectors $[c_1\,\mathbf{w}_1]_p$ and $[c_2\,\mathbf{w}_2]_p$ are nearly orthogonal for DX-k-s generators, it is harder to find c_1 and c_2 such that

$\left\| [c_1 \, \mathbf{w}_1]_p + [c_2 \, \mathbf{w}_2]_p \right\|^2$ is smaller than $\left\| [c_1 \, \mathbf{w}_1]_p \right\|^2$ or $\left\| [c_2 \, \mathbf{w}_2]_p \right\|^2$. Recall that as per our remark above, we can intuitively consider $v_{k+2}^2(k)$ as the minimum value of $\left\| [c_1 \, \mathbf{w}_1]_p + [c_2 \, \mathbf{w}_2]_p \right\|^2$ searching over $-p/2 < c_1, c_2 < p/2$; and $v_{k+1}^2(k)$ is the minimum value of $\left\| [c_1 \, \mathbf{w}_1]_p \right\|^2$ over $-p/2 < c_1 < p/2$, or, equivalently, the minimum of $\left\| [c_2 \, \mathbf{w}_2]_p \right\|^2$ over $-p/2 < c_2 < p/2$. Therefore, it is very likely that the minimizer of $\left\| [c_1 \, \mathbf{w}_1]_p + [c_2 \, \mathbf{w}_2]_p \right\|^2$ would be the same as that of $\left\| [c_1 \, \mathbf{w}_1]_p \right\|^2$ or $\left\| [c_2 \, \mathbf{w}_2]_p \right\|^2$, which gives the consistency property.

Moreover, the reason that the consistency property seems to get better with increasing s can be understood as follows. Imagine that k is much larger compared to s. As s increases, the number of coordinates in the vector $[c_1 \, \mathbf{w}_1]_p$ that is $[c_1 B]_p$ increases with s. Thus, its norm (length) also increases for a fixed c_1. The same argument goes for the vector $[c_2 \, \mathbf{w}_2]_p$. However, since the $[B]_p$ in the vectors $\mathbf{w}_1$, $\mathbf{w}_2$ are well spaced (separated by lots of zeros), taking the dot product of $[c_1 \, \mathbf{w}_1]_p$, $[c_2 \, \mathbf{w}_2]_p$ always evaluates only to $-c_1 [c_2 B]_p$. Thus, since the angle between $[c_1 \, \mathbf{w}_1]_p$, $[c_2 \, \mathbf{w}_2]_p$ is determined by

$$\cos(\theta) = \frac{[c_1 \, \mathbf{w}_1]_p' \, [c_2 \, \mathbf{w}_2]_p}{\left\| [c_1 \, \mathbf{w}_1]_p \right\| \left\| [c_2 \, \mathbf{w}_2]_p \right\|}$$

the angle between them only gets closer to $\pi/2$ with the increasing norms of $[c_1 \, \mathbf{w}_1]_p$, $[c_2 \, \mathbf{w}_2]_p$. Thus, there is even a greater chance that the solution to the minimization problem remains stable, as explained above.

The consistency property for DX-k-s generators extends from $v_{k+2}^2(k)$ to $v_{k+3}^2(k)$, with 100% consistency observed across 5000 sampled B values. This property is expected to hold for higher dimensions beyond k until there is a significant change in the relative angles among the vectors involved. FMRG-k generators also show this consistency.

The spectral test value $v_{k+d}^2(k)$ for MRGs of order k is non-increasing as d increases, meaning that the consistency property is beneficial. It ensures that the spectral value does not decline in higher dimensions, allowing for a reliable ranking of MRGs with the same order based on their spectral test values. This avoids the problem where an MRG with the best spectral value in dimension $k + 1$ might perform worse in dimension $k + 2$.

The next section focuses on finding better FMRG-k and DX-k-s generators by screening multipliers B that exhibit the consistency property and have a spectral test value below a specified threshold.

7.4   Lists of Better FMRG-K and DX-K-S Generators

Recent studies have identified several large-order FMRG-k and DX-k-s generators, with detailed findings available in cited papers. For the modulus $p = 2^{31} - 1$, specific values of k (up to 20,897) were identified for DX generators with known factorization of $p^k - 1$. With

Table 7.1 List of FMRG-k and DX-k-2 with $B < 2^{20}$ and $B < 2^{30}$ along with their spectral distance $d_{k+1}(k)$

k	p	FMRG-k				DX-k-2			
		$B < 2^{20}$	$d_{k+1}(k)$	$B < 2^{30}$	$d_{k+1}(k)$	$B < 2^{20}$	$d_{k+1}(k)$	$B < 2^{30}$	$d_{k+1}(k)$
47	2147483647	1047527	1.87e–05	1073719468	2.24e–05	1047104	1.98e–05	1073718369	1.90e–05
643	2147483647	1047252	1.91e–05	1073718413	2.14e–05	1048207	2.00e–05	1073690591	1.91e–05
1597	2147483647	1016724	2.22e–05	1073716054	1.94e–05	777051	1.90e–05	1073718474	1.97e–05
7499	2147483647	967501	1.91e–05	1073708416	2.16e–05	336838	1.87e–05	1073381825	1.84e–05
20897	2147483647	820866	2.17e–05	1073616009	2.75e–05	1028880	2.57e–05	1073658320	2.28e–05
101	2147400803	1047864	2.16e–05	1073678105	2.24e–05	1048093	1.90e–05	1073677156	1.99e–05
211	2146642319	1047749	2.02e–05	1073721711	1.97e–05	1044526	1.83e–05	1073719650	1.81e–05
307	2147431103	1045846	1.87e–05	1073691791	2.10e–05	1047799	1.92e–05	1073692149	1.99e–05
401	2147426459	1048196	2.14e–05	1073689547	2.11e–05	1033380	1.92e–05	1073689977	1.91e–05
503	2147309159	1048331	2.14e–05	1073720383	2.12e–05	1022698	1.79e–05	1073679109	1.84e–05
601	2146156163	1043822	1.91e–05	1073719452	1.92e–05	1042172	2.00e–05	1073712188	1.96e–05
701	2147262983	1046874	1.86e–05	1073717689	2.21e–05	1040476	1.89e–05	1073680615	1.82e–05
809	2145472859	1043034	2.08e–05	1073719786	2.08e–05	985209	1.89e–05	1073706375	2.00e–05
907	2143082759	1046522	1.83e–05	1073721128	2.14e–05	1032051	1.98e–05	1073718871	1.92e–05
1009	2145114779	1045259	2.15e–05	1073716427	1.99e–05	1014011	1.95e–05	1073683017	1.98e–05
1103	2140167287	1045590	2.12e–05	1073708831	2.04e–05	1028217	1.91e–05	1073718798	1.93e–05
1201	2146369943	1044395	2.16e–05	1073711360	2.18e–05	1034308	1.86e–05	1073634611	1.96e–05
1301	2146412747	1038695	2.08e–05	1073714389	2.03e–05	1019650	1.91e–05	1073694173	1.84e–05
1409	2143163459	1023942	1.89e–05	1073721338	1.93e–05	1015803	1.89e–05	1073653955	2.00e–05
1511	2144712443	1027601	1.91e–05	1073703279	2.16e–05	1033442	1.99e–05	1073711638	1.96e–05
1601	2147114687	1048172	2.15e–05	1073706909	1.93e–05	1030332	1.93e–05	1073631425	1.89e–05
1709	2146451207	1037856	1.96e–05	1073711506	2.17e–05	1001582	1.99e–05	1073379700	1.99e–05
1801	2141694407	1026509	2.22e–05	1073702501	1.97e–05	1040074	1.99e–05	1073688661	1.93e–05
1901	2147216327	1047198	1.92e–05	1073719969	1.76e–05	1002056	1.88e–05	1073576432	1.95e–05
2003	2147438687	1040900	1.95e–05	1073694634	2.02e–05	964935	1.86e–05	1073695597	1.84e–05
2111	2143947263	1048318	1.90e–05	1073719627	1.96e–05	1023929	1.85e–05	1073516503	1.84e–05
2203	2141440559	1041675	1.89e–05	1073709809	1.77e–05	1034145	1.88e–05	1073687710	1.95e–05
2309	2147143463	1046953	1.76e–05	1073719650	2.16e–05	980142	1.79e–05	1073716624	1.91e–05
2411	2138227199	1046643	2.05e–05	1073720277	2.02e–05	960701	1.91e–05	1073681521	1.84e–05
2503	2133944399	1045562	2.10e–05	1073719199	1.90e–05	1039452	1.87e–05	1073702140	1.92e–05
2609	2138671967	1032235	2.15e–05	1073719690	1.91e–05	921424	1.94e–05	1073719923	1.89e–05
2707	2146370063	1048221	1.80e–05	1073697994	2.19e–05	923007	2.00e–05	1073642864	1.99e–05
2801	2146388039	1040902	2.07e–05	1073721289	1.87e–05	1029014	1.87e–05	1073715771	1.98e–05
2903	2133427823	1048504	1.97e–05	1073720598	1.88e–05	620667	2.00e–05	1073655527	1.89e–05
3001	2144425247	1048008	2.18e–05	1073693367	2.24e–05	973895	1.90e–05	1073531231	1.85e–05
3109	2140742519	1028657	1.88e–05	1073702701	2.14e–05	842603	1.91e–05	1073609758	1.98e–05
3203	2142764759	1038126	2.19e–05	1073707984	2.19e–05	1045174	1.83e–05	1073494301	1.87e–05
3301	2132602463	1029102	1.98e–05	1073700259	2.21e–05	927480	1.94e–05	1073599925	1.84e–05
3407	2141240639	1025440	1.83e–05	1073720463	1.85e–05	1036658	1.82e–05	1073570507	1.97e–05
3511	2146070687	1044201	1.76e–05	1073699334	1.87e–05	931391	1.82e–05	1073696749	1.96e–05

(continued)

Table 7.1 (continued)

		FMRG-k				DX-k-2			
k	p	$B < 2^{20}$	$d_{k+1}(k)$	$B < 2^{30}$	$d_{k+1}(k)$	$B < 2^{20}$	$d_{k+1}(k)$	$B < 2^{30}$	$d_{k+1}(k)$
3607	2146457063	1016279	2.02e–05	1073716724	1.77e–05	1000578	2.00e–05	1073593347	1.98e–05
3701	2135907023	1025699	2.21e–05	1073715028	2.00e–05	1002227	1.98e–05	1073619234	1.86e–05
3803	2115425519	1012956	1.78e–05	1073713780	2.07e–05	1044969	1.96e–05	1073715220	1.89e–05
3907	2130101999	1017573	1.88e–05	1073697705	1.84e–05	980525	1.96e–05	1073545500	1.89e–05
4001	2143071167	1044560	1.85e–05	1073666248	2.18e–05	694433	1.98e–05	1073431583	1.99e–05
5003	2146224359	1033689	1.81e–05	1073727083	2.11e–05	1033485	2.47e–05	1073741516	2.14e–05
6007	2137498943	1035429	2.16e–05	1073737595	1.94e–05	1015366	1.84e–05	1073682415	1.87e–05
7001	2146873559	1010706	2.15e–05	1073709808	1.79e–05	954617	2.69e–05	1073720260	2.30e–05
8009	2142326903	1041446	2.05e–05	1073717208	2.46e–05	960779	2.18e–05	1073724902	2.35e–05
9001	2140247399	1045508	2.37e–05	1073737583	2.44e–05	1029903	2.37e–05	1073682825	2.30e–05
10007	2147051903	954436	1.90e–05	1073688561	2.09e–05	987064	2.15e–05	1073702542	2.17e–05
11003	2146207223	1006998	1.93e–05	1073664067	1.81e–05	1047362	2.74e–05	1073730907	2.47e–05
12007	2109950867	990386	2.75e–05	1073714070	1.93e–05	879930	2.13e–05	1073692289	2.41e–05
13001	2147191153	1034426	2.06e–05	1073566489	1.91e–05	827268	2.06e–05	1073621150	2.34e–05
14009	2146857347	1046516	2.66e–05	1073730141	2.74e–05	1041819	2.20e–05	1073738204	2.13e–05
15013	2138487383	1030728	1.98e–05	1073724930	1.85e–05	993180	2.13e–05	1073723417	1.93e–05
20011	2121351707	938904	2.32e–05	1073528611	2.44e–05	687847	2.24e–05	1073460800	2.11e–05
25013	2135944739	1007372	1.91e–05	1073707771	2.06e–05	969323	2.15e–05	1073692717	2.56e–05

a modulus of the form $p = 2^{31} - w$, generators with k up to 25,013 were provided. The focus was on finding generators with the maximum period, although some may have poor spectral test values in higher dimensions.

Using a spectral distance algorithm, a new set of generators with improved spectral distance $d_{k+1}(k)$ was found. The search prioritized multipliers with spectral distances below a specified bound, then checked for consistency and maximum period. Table 7.1 lists FMRG-k and DX-k-2 generators with smaller spectral distances, indicating a good lattice structure. For efficiency, the spectral distances were precisely calculated using the LLL algorithm, and exact values were verified.

Part III

Parallelization of Random Number Generators

Automatic Generation Method for Parallelization of LCG and DX Generators

A parallel Pseudo-Random Number Generator (parallel PRNG) is a method to generate several parallel streams of pseudo-random numbers. Such parallel streams are usually deployed or used by several processors working in parallel on some computing task. There are many desirable properties that such methods should have:

1. **Independence**: It is desirable that streams generated across processors have some "independence". Operationally, this means that the stream on any processor must not be an "easy" transform of the stream on another. They will be related, but the relation should not be a simple computational transform element by element.
2. **Reproducibility**: The PRNG is initialized with a seed, and a desirable property is that parallel sequences remain the same for the same seed; this ensures that experiments are repeatable.
3. **Portability**: The generated sequences should not depend on the hardware or software platform.
4. **Efficiency**: Generation of a next number should be fast as well as initializing a new sequence should be efficient.
5. **Long period**: The parallel sequences should have long periods. This is a desirable property of any PRNG, and it should be true of a PRNG too.
6. **Scalable**: The parallel PRNG should ideally be able to extend to many processors, i.e., be able to generate many parallel sequences with the above properties.

We first review classical parallelization methods for LCGs and describe the newly proposed automatic generation method. We then discuss possible extensions to a class of efficient larger-order MRGs including DX generators.

L. Deng et al., *Random Number Generators for Computer Simulation and Cyber Security*, Synthesis Lectures on Mathematics & Statistics, https://doi.org/10.1007/978-3-031-76722-7_8

8.1   Classical Methods

We begin this discussion with the parallelization of LCG, the best-known uniform random number generator (see Chap. 2):

$$X_i = BX_{i-1} + A \bmod p, \quad i \geq 1, \quad U_i = X_i/p$$

It is common to set additive constant $A = 0$. When p is a prime, maximum period length is $p - 1$, and B is a *primitive root* mod p. For example, the most popular LCG is with multiplier $B = 16807 = 7^5$ and modulus $p = 2^{31} - 1$ with period length $p - 1 = 2147483646 \approx 2.1 \times 10^9$. Note $B = 7$ is the smallest primitive root for p but it is too small to be a "good" multiplier for the LCG. See the discussion on statistical justification in Chap. 6.

The parallelization of LCG can be achieved with different disjoint sequences in several ways. The most common ones are discussed in Deng et al. (1994) as follows:

1. **Single starting seed**. Assign one process to generate a stream of random numbers and all other processes obtain random numbers from this single stream. This method has several difficulties. First, it is difficult to balance the speed of production and consumption. Second, if the processes are not completely synchronized in their requests for random numbers, then the assignment of random numbers to processes may not be reproducible. Third, in the case of message-passing systems, the communication time required may be several orders of magnitude greater than that required for generating random numbers locally.

   into a finite number, n, of subsequences with very far apart starting points. Each process is assigned a nonoverlapping subsequence, the starting point of which can easily be pre-computed easily. Clearly, the sequences used by all processes belong to the same LCG sequence with B is a primitive root and they are (long) shifts of each other. In addition, the number of nonoverlapping pseudo-random variates is only $(p - 1)/n$. There are at most n different processes with "independent" random number sequences.

Most common methods divide LCG sequence into several subsequences with *different* starting seed. Therefore, LCG for each processor has *same multiplier*. However, they also shared the same problem: since LCG has short period length (2×10^9), hard to avoid overlap among subsequences. This is a violation of the key assumption of *independence* among different processors. There are three common methods with different starting seeds as discussed below:

1. **Fixed block size method**. For the fixed block size method, each subsequence (row) is an LCG with the same B and a new starting seed $X_{new} = B^n X_{old} \bmod p$. We can *pre-compute* and *pre-store* $D = B^n \bmod p$ using a quick method to jump n steps ahead because we can use an efficient algorithm (of $O(log_2(n))$) to compute $B^n \bmod p$ as shown below:

RNG1	X_0	X_1	X_2	$\cdots$	X_{n-2}	X_{n-1}
RNG2	X_n	X_{n+1}	X_{n+2}	$\cdots$	X_{2n-2}	X_{2n-1}
RNG3	X_{2n}	X_{2n+1}	X_{2n+2}	$\cdots$	X_{3n-2}	X_{3n-1}
$\vdots$	$\vdots$	$\vdots$	$\vdots$	$\cdots$	$\vdots$	$\vdots$

2. **Fixed step leapfrog**. The LCG sequence is divided by *leapfrogging* (fixed) n steps and each subsequence (row) is another LCG with the same multiplier, B^n mod p as shown below:

RNG1	X_0	X_n	X_{2n}	X_{3n}	$\cdots$ $\cdots$
RNG2	X_1	X_{n+1}	X_{2n+1}	X_{3n+1}	$\cdots$ $\cdots$
RNG3	X_2	X_{n+2}	X_{2n+2}	X_{3n+2}	$\cdots$ $\cdots$
$\vdots$	$\vdots$	$\vdots$	$\vdots$	$\vdots$	$\cdots$ $\cdots$

We need to choose step size n carefully because maximum period can be achieved *only when* the step size n satisfies $\gcd(n, p - 1) = 1$.

3. **Lehmer tree**. This is a popular method proposed by Frederickson et al. (1984) where two LCGs are used:

   a. $L(X) = B_L X$ mod p (use LCG with multiplier B_L to compute a new starting seed)
   b. $R(X) = B_R X$ mod p (use LCG with multiplier B_R for all processors).

This method is not better than others mentioned because it shares the same problems. Both LCGs will have the *same* LCG sequence with "random" starting seed. However, "random" division of the LCG sequence will yield an even *shorter* disjoint stream. The random starting point method is similar to the method of predetermined starting point except that the starting seed for each process is computed "randomly." Because the same multiplier is used in all sequences, they again belong to the same class. Therefore, they are ("random") shifts of each other. The expected shortest disjoint stream can be shown to be $(p - 1)/n^2$, where n is the number of random starting points used.

8.2   Leap Frog Method for LCG

The *systematic random leap frog method* for LCG was proposed by Deng et al. (1994) with several key components below:

1. Let B be a multiplier of LCG which is a *primitive root* mod p.
2. Each new processor is assigned to an LCG with *different multiplier* as:

   a. use *another LCG* to select leapfrog steps (r's),
   b. compute the multiplier based on the step size, B^r mod p.
   c. *all generated LCGs* automatically have the period $(p - 1)$.

8.2.1   Systematic Random Leap Frog Method

Deng et al. (1994) proposed to generate a sequence of random numbers using an LCG such that every generated number r is relatively prime to $p - 1$ for p chosen to maximize the period of the LCG. The method is described as: first choose a multiplier, R, that is a primitive element modulo $p - 1$ and is relatively prime to $p - 1$; then, generate a sequence of such r's using recursive equation:

$$r_n = R\, r_{n-1} \bmod (p - 1), \ n \geq 1, \tag{8.1}$$

where r_0 is any non-zero initial seed that is relatively prime to $p - 1$; $r_0 = 1$ for simplicity. Using Eq. 8.1, distinct r_n's (within the period of the LCG in Eq. 8.1) can be produced that are relatively prime to $p - 1$. These values can be used to assign each processor in a parallel environment its own LCG using the systematic leapfrog method proposed by Deng et al. (1994). The algorithm for the systematic leapfrog is summarized in Algorithm 8.1:

Algorithm 8.1: Algorithm for LCG systematic leapfrog

Suppose B and R are the multipliers of the LCG for the first processor and the baseline LCG in Eq. 8.1, respectively.

1. When initiating a new processor, generate a new r using the following equation:

   $$r_{new} = R\, r_{old} \bmod (p - 1).$$

2. Calculate a multiplier B_{new} for the new processor as

   $$B_{new} = B^{r_{new}} \bmod p.$$

3. The new processor is assigned LCG with multiplier B_{new} as

   $$X_i = B_{new} X_{i-1} \bmod p, \ i \geq 0.$$

Systematic Leap Frog: Initial Setup

1. Initial conditions: For a given prime modulus, p
   a. Choose a multiplier B of LCG which is a primitive root mod p.
   b. Choose R which is a multiplier of LCG with maximal period with modulus $(p-1)$.
   c. Choose $\gcd(r_0, p-1) = \gcd(R, p-1) = 1$. We can simply choose initial $r_0 = 1$.

2. The *key idea* of the proposed method is touse multiplier R of the LCG to generate the "random" leap step size r_j that $\gcd(r_j, p-1) = 1$ for all $j = 0, 1, \ldots$.

Systematic Leap Frog: Selection of p

1. We use LCG with (composite) modulus $(p-1)$:

$$r_j = R r_{j-1} \bmod (p-1), \quad j \geq 1.$$

2. Within its LCG period,
   a. different leap sizes (multipliers) are automatically (and randomly) generated.

3. We choose R and p with longer period
   a. choose p so that $Q = (p-1)/2$ is also a prime.
   b. e.g. $p = 23$, $Q = (23-1)/2 = 11$.
   c. choose R to achieve the maximum period of $Q - 1$. Such R is called *primitive element* mod $(p-1)$.

Next, we demonstrate the new algorithm with an LCG with a small modulus.

Example 8.1 (*Systematic leap frog: Illustration*)

Recall that LCG with $p = 23$, maximum period with $B = 5, 7, 10, 11, 14, 15, 17, 19, 20, 21$ and *all* of its eight primitive roots are automatically generated from the above method (Table 8.1). For more details on the above methods mentioned, (see Deng et al. 1994; Mascagni 1998; Mascagni and Srinivasan 2000; L'Ecuyer et al. 2002).

Here we mention a few different techniques for the case of different starting seeds which does not store them pre-computed, as that is not scalable. Then, we discuss the last method above in more detail where a technique called *systematic leapfrog method* proposed by Deng et al. (1994) to automatically choose different multipliers for the corresponding LCGs so that they are of the maximum period.

For many of the parallel PRNGs, a fixed block method is used to assign substreams to separate processors. This requires an efficient means of skipping ahead in the generator sequence to provide adjacent segments that do not overlap.

Table 8.1 Systematic leap frog with $B = 5$, $p = 23$, and $R = 7$

i	Step 1 r	Step 2 B^r mod p	Step 3 $r_{new} =$ $R\,r_{old}$ mod $(p - 1)$
1	1	5	7
2	7	17	5
3	5	20	13
4	13	21	3
5	3	10	21
6	21	14	15
7	15	19	17
8	17	15	9
9	9	11	19
10	19	7	1 (repeat)

Example 8.2 (*LCG systematic leapfrog correlation effect with p = 107*)

To show the effect of correlation among various processors, we choose LCG(B, p) with $B = 5$ and $p = 107$. For the fixed block method, we choose five initial seeds X_0 so that they are roughly far apart from each other. For the systematic leapfrog method, we choose five different multipliers B each resulting in LCGs with the maximum period. The pairwise scatter plots for the two methods are shown in Fig. 8.1.

From the left panel of Fig. 8.1, one can see that there are clear patterns of linear relations between the variates generated from different streams for the fixed block method. This problem of a high correlation among different streams was first reported in Anderson and Titterington (1993). On the other hand, no patterns can be observed among different streams for the systematic leap frog method as shown in the right panel of Fig. 8.1.

8.2.2   Advantage of Systematic Leap Frog Method

The "systematic leap frog method" is superior to the "fixed block method" for specifying PRNGs for different processors because under the leapfrog method, each processor has its own LCG so the potential overlapping problem of the generated LCG sequences does not exist. Furthermore, the correlation and linear structure problems are greatly minimized as shown in Fig. 8.1. While the leapfrog method minimizes the problem of "within processor correlation" with different LCGs for different processors, the problem of "within processor correlation" (with the same LCG) may still exist however. Next, we discuss a natural way

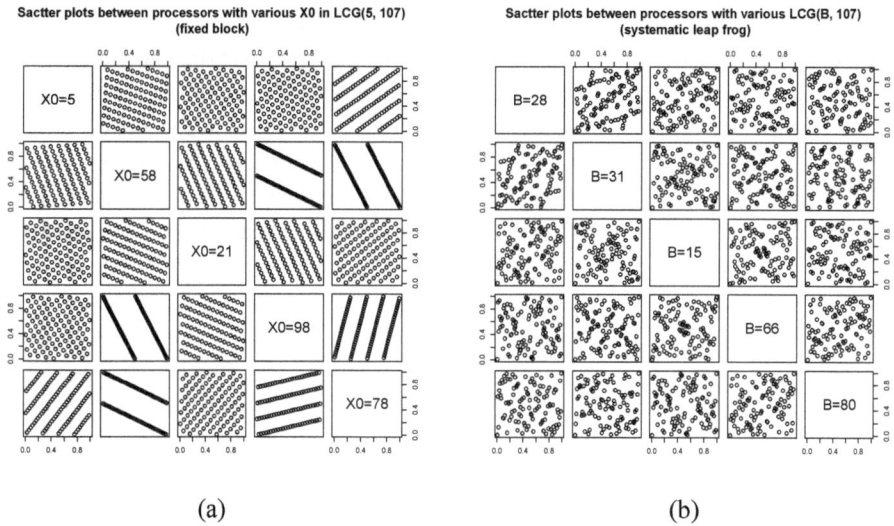

(a)   (b)

Fig. 8.1 Scatter plots of correlation between pairs of five processors. The left panel is for the fixed block method and the right panel is for the systematic leap frog method

to minimize the within-processor correlation by replacing the LCG with a large-order MRG as the baseline generator.

It was pointed out above that random number sequences produced with a fixed multiplier and different starting seeds were shifts of each other. If X and Y are maximum cycle RNGs sharing a common multiplier B, then there is a positive integer m such that

$$\{Y_0, Y_1, Y_2, ...\} = B^m \{X_0, X_1, X_2, ...\} \bmod \ p.$$

If $B^m \bmod \ p$ happens to be a small number, then the two sequences will be "parallel" to each other. This problem is more pronounced for sequences of short to moderate length. In contrast, the method proposed in the next section assigns different Bs to each processor, and the generated sequences are not "parallel" to each other.

Consider next the problem of robustness. It is generally accepted that a "good" LCG should have the maximum period and easily pass the spectral test. However, it is also understood that even the "best" pseudo-random number generator may fail some (perhaps unknown) empirical tests and perform poorly in some applications. If we use the same multiplier for all applications, it may fail miserably in some particular applications. In contrast, methods using different multipliers should be more robust because they do not rely on the performance of just one RNG. Intuitively, to run a simulation 1,000 times, it is better to use a distinct RNG with a unique multiplier for each run of the simulation. It is more robust because the whole simulation will not be seriously affected by a small number of "bad" RNGs. The systematic random leapfrog method proposed in the next section chooses different Bs for different RNGs; this reduces the likelihood of such dramatic failure.

Let us conclude this section by considering similar problems in uniprocessor systems. The need to provide random number sequences in multiprocessor systems focuses our attention on the quality of a set of random number sequences. The core of the problem is: There are different processes requiring independent sequences of random numbers. We need to ensure that each sequence is indistinguishable from that produced by a random sequence of standard uniformly distributed random variables; and that a sequence generated for one process is independent of the sequence generated for another process. The question of the location of the processes is relatively minor because the same problem is faced in uniprocessor systems. Consider, for example, the case of large-scale discrete system simulations. Typically, a complex system is analyzed into a large number of interacting subsystems. Different random events in the system, such as interarrival times and service times for different servers, need different sequences of random numbers. Simulation packages, such as GPSS, provide a small number of random number generators, and different random events have to get their random numbers from these generators. There is no guarantee that the sequences of random numbers are supplied to different random events. In fact, the quality of the random number sequences depends on accidental features of the program, such as the order of statements, and initial conditions of the simulations. The reliability of simulations on uniprocessor systems would be much improved if each random event was assigned a different RNG.

8.3   Automatic Generating Method for Parallel MRGs (AGM-1)

We will discuss the parallelization of DX generators proposed by Deng and Xu (2003) using a similar idea to the random leap-frog method for LCG as proposed by Deng et al. (1994).

8.3.1   Choosing Different Seeds for MRGs

Let

$$
\mathbf{M}_f = \begin{pmatrix}
0 & 1 & 0 & \ldots & 0 \\
0 & 0 & 1 & \ldots & 0 \\
0 & 0 & 0 & \ldots & 0 \\
. & . & & \ldots & . \\
& & & \ldots & \\
0 & 0 & 0 & \ldots & 1 \\
\alpha_k & \alpha_{k-1} & . & . & \alpha_1
\end{pmatrix}
$$

be the companion matrix of the k-th order MRG. We can then generate the k-dimensional seed vector that is "d-apart" from the initial seed vector $\mathbf{X}_{old} = (X_0, X_1, \ldots, X_{k-1})'$,

$$\mathbf{X}_{new} = \mathbf{M}_f^d \, \mathbf{X}_{old} \bmod p.$$

1. When $k = 1$, it is the efficient jump ahead scheme with $\mathbf{M}_f^d = B^d \bmod p$ which can pre-stored for jumping d steps ahead.
2. When k is small, one can quickly compute the $k \times k$ matrix $\mathbf{M}_f^d$ and pre-store that matrix.
3. When k is large, this jumping ahead scheme is very inefficient.

One way is to borrow the idea from the Systematic Random Leap Frog Method for LCG Deng et al. (1994). We describe such method, called *Automatic Generation Method (AGM)* for MRG as proposed in Deng (2004) as described next.

8.3.2   Theory Behind Construction Method

Theorem 8.1   (Deng (2004))

- Let $R(k, p) = (p^k - 1)/(p - 1)$ be a prime and c be any non-zero integer.
- Assume we found one characteristic polynomial

$$f(x) = x^k - \alpha_1 x^{k-1} - \dots - \alpha_{k-1} x - \alpha_k$$

to be a primitive polynomial.

We can apply the theory and define

$$G(x) = c^{-k} f(cx) = x^k - \sum_{i=1}^{k} G_i x^{k-i} \bmod p$$

$$H(x) = -\alpha_k^{-1} x^k f(c/x) = x^k - \sum_{i=1}^{k} H_i x^{k-i} \bmod p$$

where

$$G_j = c^{-j} \alpha_j \bmod p$$

and

$$H_j = -\alpha_k^{-1} \alpha_{k-j} c^j \bmod p,$$

for $j = 1, 2, \dots, k$, $\alpha_0 = -1$. If $G_k \, (= c^{-k} \alpha_k)$ is a primitive root mod p, then both $G(x)$ and $H(x)$ are primitive polynomials.

Note that MRGs corresponding to $G(x)$ and $H(x)$ will have maximal period if c is chosen carefully to satisfy the condition given. Most importantly, there are *same number* of nonzero

Table 8.2 Coefficients of DX-k-s

	G_1	G_k	H_{k-1}	H_k
$s = 1$	c^{-1}	$c^{-k}B$	$-c^{k-1}B^{-1}$	c^kB^{-1}
$s = 2$	$c^{-1}B$	$c^{-k}B$	$-c^{k-1}$	c^kB^{-1}

terms in the corresponding MRGs so that corresponding MRGs are efficient. Furthermore, many such maximal period MRGs can be found quickly by changing the values of c.

8.3.3   Construction of MRG-k-s from DX-k-s

As mentioned previously, only s non-zero coefficients of $G(x)$, $H(x)$ from DX-k-s and corresponding MRG-k-s (MRG with s non-zero terms) is reasonably efficient. In particular, for $s = 1, 2$, we can find their coefficients from Table 8.2.

8.3.4   Key Ideas Behind AGM

Listed below are some key ideas for the AGM algorithm:
1. Since B is a primitive root, c can be written as $c = B^d \bmod p$, for some d.
2. $c^{-k}B = B^{(-kd+1)} \bmod p$ is a primitive root, if any only if $\gcd(kd - 1, p - 1) = 1$.
3. Borrowing the procedure for the systematic leapfrog method, we can generate a sequence of r_i so that $\gcd(r_i, p - 1) = 1$. Hence, one can then solve

$$d = k^{-1}(r_i + 1) \bmod (p - 1).$$

8.4   Parallelization of DX Generators

Design Consideration
1. For efficiency, choose DX-k-s with $s = 1$ or $s = 2$
2. For a prime k, find $p < 2^{31}$ [32-bit RNGs] such that $(p^k - 1)/(p - 1)$ is a GMP
3. In addition: $Q = (p - 1)/2$ is a prime.
   a. To maximize its period of $r_j = R \ r_{j-1} \bmod (p - 1)$
   b. For the existence of $k^{-1} \bmod (p - 1)$

Initial Setup

1. Let B be a multiplier for DX-k-s RNG.
2. $f(x)$ is the corresponding primitive polynomial for DX-k-s.
   a. $s = 1$: $f(x) = x^k - x^{k-1} - B$
   b. $s = 2$: $f(x) = x^k - Bx^{k-1} - B$
3. R is a multiplier with maximal period for LCG under mod $(p - 1)$.
4. $\gcd(r_0, p - 1) = 1$ and $\gcd(R, p - 1) = 1$.
5. We can simply choose $r_0 = 1$.

General Algorithm

A sequence of different MRGs with maximal period corresponding to $G(x)$ or $H(x)$ can be constructed:

1. $r_j = Rr_{j-1} \bmod (p - 1)$.
2. $d_j = k^{-1}(r_j + 1) \bmod (p - 1)$.
3. $c_j = B^{d_j} \bmod p$.
4. Compute primitive polynomials

$$G(x) = c_j^{-k} f(c_j x) \bmod p$$

and

$$H(x) = -x^k f(c_j/x)/B \bmod p.$$

Let's consider the following example which takes $k = 4001$ for DX-k-2.

Example 8.3 (*Illustration for DX-4001-2*)

DX-4001-2 generator is

$$X_i = 1031978(X_{i-1} + X_{i-4001}) \quad \bmod\ 2143071167$$

has the period length $\approx 10^{37333.5}$.

The question is how to parallelize such a DX generator. As mentioned previously, for PRNG using large order MRG (e.g. DX-4001-2), changing seeds far apart for an MRG is *inefficient* when the order k is large.

Initial Steps

1. For $k = 4001$.
2. Find $p = 2^{31} - 4412481 = 2143071166$

    a. $(p - 1)/2$ is a prime.
    a. $(p^{4001} - 1)/(p - 1)$ is a GMP.

3. For $s = 2$, find DX-4001-2 with $B = 1031978$.
4. Find $R = 33455$, as a multiplier with maximal period in LCG (with $r_0 = 1$)

$$r_j = R\,r_{j-1} \quad \text{mod} \ (p - 1)$$

First Iteration

1. Start with $r_0 = 1$, find next

$$r_j = 33455 r_{j-1} \text{ mod } (p - 1)$$

2. $r_1 = 33455$
3. $d_1 = k^{-1}(r_j + 1) \text{ mod } (p - 1) = 1576906160$
4. $c_1 = B^{d_1} \text{ mod } p = 271596069$.
5. $G_1 = c_1^{-1} B \text{ mod } p = 538038547$,
6. $G_{4001} = c_1^{-4001} B \text{ mod } p = 466567840$.

$$r_2 = 33455^2 \text{ mod } (p - 1) = 1119237025$$

Iteration Results

The iteration results are tabulated in Table 8.3.

PRNG for MRG: Generated MRGs

For $k = 4001$ and $p = 2143071167$, we have $G_1 = 538038547$, $G_{4001} = 466567840$. The corresponding MRG-4001-2 generator

$$X_i = 538038547 X_{i-1} + 466567840 X_{i-4001} \text{ mod } 2143071167$$

has the period of $10^{37333.5}$. In contrast, the period length for classical LCG is $\approx 2 \times 10^9$ and period length of MT19937 is $\approx 10^{6001}$.

Highlights

The number of maximal period MRGs can be generated with $k = 4001$ is 1,071,535,582 each with maximal period of $10^{37,333}$. The speed of production is more than 100,000 MRGs/second whereas the initial search of DX-4001 may take 1–2 days.

    Deng (2004) proposed an efficient method for finding large-order, maximum-period MRGs. He also described a method for constructing maximum-period MRGs from a single MRG for DX generators in particular. Denote by MRG-k-s the class of maximum-period MRGs of order k with s nonzero coefficients in their corresponding characteristic polynomials. When $R(k, p) - (p^k - 1)/(p - 1)$ is a prime, Deng (2004), Deng et al. (2009a)

Table 8.3 List of the first 20 iterations of AGM algorithm for constructing MRG-4001-2

n	r_n	c_n	G_1	G_{4001}	H_{4000}	H_{4001}
1	33455	271596069	538038547	466567840	377755423	784137450
2	1119237025	869504607	550884537	478847729	657202932	1753090457
3	335259023	442515096	1566662175	187227285	1296770865	1857614561
4	1399202787	104753893	1315679652	1107629070	810654320	328178428
5	1368831313	986411888	1651288829	254685660	1154006112	1220729562
6	1106901327	173373102	2075793756	1281741128	1384090581	906013825
7	1257217471	379485739	1845836277	409133161	1335004469	1908485007
8	295788389	1096956795	1026567566	47651946	1418866529	1884598511
9	1040980573	33287558	1992756369	815795955	1878987105	518749096
10	1098622215	948019578	1188384449	237670234	1538951230	966296673
11	735705925	1736486493	1922179869	783700166	1125176950	651066012
12	2012450531	1743691697	2093852176	1148153589	2087924570	1875568645
13	1951834715	1940817358	1962052560	788843490	1120647280	496161642
14	1395033471	1164585786	1000823826	2090072189	1793563909	1630879776
15	1183990323	1242931911	1134617119	1105902684	35171343	41708201
16	11894787	1788163231	143622906	356341728	1144623554	2016376913
17	1471933375	26911583	1391019115	1651184672	1885722755	2126035671
18	41808277	1136237452	721007986	805583938	2123397604	2075579134
19	1413506803	348179413	1501753337	1577403115	939249849	823191446
20	2004816575	1461887585	628558650	838743597	614541364	1013965539

proposed a simple method to construct MRG-k-s generators from the characteristic polynomial of a DX-k-s generator. Using an idea similar to the systematic leap frog method, Deng (2004) proposed an automatic generating method (AGM-1) for finding values of c such that

$$G(x) = c^{-k} f(cx) \bmod p \qquad (8.2)$$

is a primitive polynomial, where $f(x)$ is the characteristic polynomial of the MRG-k-s as in Eq. 4.1. The new characteristic equation in Eq. 8.2 can in turn define an MRG with a different structure from the MRG defined by $f(x)$ (see Deng 2004; Deng et al. 2009a, for more details).

Algorithm AGM

The algorithm AGM Deng and Shiau (2015) is discussed below.

Algorithm 8.2: Automatic Generating Method (AGM)

Let R be a primitive element modulo $p - 1$ such that $\gcd(R, p - 1) = 1$. The following procedure will randomly generate a sequence of maximum-period MRGs.

1. When a new processor is initiated, generate r_n by the following equation:

$$r_n = R\, r_{n-1} \bmod (p - 1), n \geq 1 \text{ with } r_0 = 1.$$

2. Calculate d_n and c_n for the new processor by

$$d_n = k^{-1}(r_n + 1) \bmod (p - 1) \text{ and } c_n = B^{d_n} \bmod p.$$

3. With the given c_n, compute the primitive polynomial $G(x)$ by

$$G(x) = c_n^{-k} f(c_n x) \bmod p.$$

4. The new processor is assigned the newly constructed maximum-period MRG corresponding to the characteristic polynomial $G(x)$ as follows:

$$X_i = G_1 X_{i-1} + \cdots + G_k X_{i-k} \bmod p,$$

Using the Algorithm AGM many MRGs (up to 100,000 per second) with maximum period may be quickly found. Each of the MRGs found can be assigned to one of the subtasks or processors in a parallel simulation. Deng et al. (2009b) extended this approach to generate different MRGs "randomly," quickly, and automatically while maintaining the maximum-period property.

To perform computational tasks in parallel, each of a number of processors must be able to generate a unique stream of pseudo-random numbers that are independent of streams from other processors. Otherwise, even if each processor is given a unique starting seed, there would be a chance for streams generated by any two processors to overlap at some point. This is particularly true if the period length of the PRNG is small or the number of multiprocessors is large. Overlapping streams of random numbers would introduce the potential for long-range correlation between different processors, jeopardizing the validity of simulations or Monte Carlo results.

We believe that it is a good idea to use different MRGs (via AGM Algorithm) for different processors so that we can minimize the possible (unknown) "defects" on a fixed generator for certain applications.

Suppose we wanted to conduct a coin-tossing experiment to generate 1,000,000 bits. In that case, we have two options: either select one (hopefully unbiased) coin and toss it

1,000,000 times or have 1,000 people each select and toss 1,000 different coins. The first method is less advisable because it heavily depends on the assumption that the chosen coin is fair, and it would also take significantly more time to complete.

One naive method of jumping ahead is simply to multiply the generator's state space by a pre-computed matrix of a very large order. However, this operation would require excessive memory space and can be very time-consuming to perform. Using MT19937, a number of methods for "jumping ahead" in the generated sequence to obtain different starting seeds have been proposed (see, Deng et al. 2008a; Haramoto et al. 2008b). The main advantage of the automatic generating methods described earlier is that each processor can be (quickly) assigned with a distinct maximum-period MRG. Therefore, the process does not suffer from the potential problem of long-range correlation between different processors.

In this chapter, we consider a new class of general MRGs defined by the following characteristic polynomial

$$f(x) = (x - B)(x - C)^{k-1} - ABx^{k-2} \bmod p \tag{9.1}$$

where A, B, and C are suitably chosen nonzero integers over $\mathbb{Z}_p$ such that $f(x)$ is k-th order primitive polynomial modulo p. By expanding this polynomial with the help of the binomial theorem, we obtain k nonzero multipliers $\alpha_1, \alpha_2, \ldots, \alpha_k$, each of which will be a function of the parameters A, B, or C:

$$\alpha_i = \begin{cases} (k-1)C + B \bmod p, & \text{for } i = 1 \\ AB - \binom{k-1}{2}C^2 + \binom{k-1}{1}BC \bmod p, & \text{for } i = 2 \\ (-1)^{i-1}(\binom{k-1}{i}C^i + \binom{k-1}{i-1}BC^{i-1}) \bmod p, & \text{for } i = 3, 4, \ldots, k-1 \\ (-1)^{k-1}BC^{k-1} \bmod p, & \text{for } i = k \end{cases} \tag{9.2}$$

We refer to this new class of general MRGs as DW-k generators Deng et al. (2023). Since the multipliers $\alpha_1, \alpha_2, \ldots, \alpha_k$ are fully specified by the parameters A, B, C, we denote DW-k as DW$(k; A, B, C; m)$ when specifying the order k, parameters A, B, C, and prime modulus p. To avoid confusion, we refer to A, B, C as the *parameters* of the $\alpha_1, \alpha_2, \ldots, \alpha_k$ *multipliers*.

We will now show that DW-k can be implemented efficiently and in parallel using a matrix congruential generator (MCG) which shares the same characteristic polynomial as Eq. 9.1. The MCG is defined iteratively by the linear recurrence

$$\mathbf{X}_i = \mathbf{B}\mathbf{X}_{i-1} \bmod p, \quad i \geq 1, \tag{9.3}$$

where $\mathbf{X}_i$'s are k-dimensional vectors over $\mathbb{Z}_p^k$, and $\mathbf{B}$ is a $k \times k$ multiplier matrix over $\mathbb{Z}_p^{k \times k}$. The recurrence is initialized by the not-all-zero integer seed vector $\mathbf{X}_0 \in \mathbb{Z}_p^k$. The characteristic polynomial of the MCG is defined as

$$f_{\mathbf{B}}(x) = \det(x\mathbf{I} - \mathbf{B}) \bmod p. \tag{9.4}$$

As a vector sequence, MCG has the maximum period of $p^k - 1$ *if and only if* $f_{\mathbf{B}}(x)$ is a primitive polynomial. Maximum period MCGs also have the equi-distribution property up to order k and the extremely long period $p^k - 1$. A brief review of the MCG is given by L'Ecuyer (1990).

Define the multiplier matrix

$$\mathbf{B} = \begin{pmatrix} B & 0 & 0 & \dots & 0 & A \\ B & C & 0 & \dots & 0 & 0 \\ B & C & C & \dots & 0 & 0 \\ \vdots & \vdots & \vdots & \ddots & \vdots & \vdots \\ B & C & C & \dots & C & 0 \\ B & C & C & \dots & C & C \end{pmatrix}. \tag{9.5}$$

In Sect. 9.1, we will prove that the MCG's characteristic polynomial $f_{\mathbf{B}}(x)$ is equal to $f(x)$ in Eq. 9.1. The linear recursion in Eq. 9.3 can be rewritten as

$$\begin{pmatrix} X_{i,1} \\ X_{i,2} \\ X_{i,3} \\ \vdots \\ X_{i,k} \end{pmatrix} = \begin{pmatrix} BX_{i-1,1} + AX_{i-1,k} \\ BX_{i-1,1} + CX_{i-1,2} \\ BX_{i-1,1} + CX_{i-1,2} + CX_{i-1,3} \\ \vdots \\ BX_{i-1,1} + CX_{i-1,2} + \dots + CX_{i-1,k} \end{pmatrix} \bmod p, \quad i \geq 1 \tag{9.6}$$

and efficiently implemented as

$$\begin{pmatrix} X_{i,1} \\ X_{i,2} \\ X_{i,3} \\ \vdots \\ X_{i,k} \end{pmatrix} = \begin{pmatrix} BX_{i-1,1} + AX_{i-1,k} \\ BX_{i-1,1} + CX_{i-1,2} \\ X_{i,2} + CX_{i-1,3} \\ \vdots \\ X_{i,k-1} + CX_{i-1,k} \end{pmatrix} \bmod p, \quad i \geq 1. \tag{9.7}$$

Notice in Eq. 9.7 that the generated output $X_{i,1}$ and $X_{i,2}$ are solely generated from numbers in the previously generated output vector $\mathbf{X}_{i-1}$. However, $X_{i,j}$ for $j = 3, 4, \dots, k$ is the sum of the previous number just generated in the current output vector $\mathbf{X}_i$ and a multiple of a number in the previous output vector $\mathbf{X}_{i-1}$. Hence, the recursion in Eq. 9.7 does not require as many multiplications and additions as the one in Eq. 9.6. Also, we note that additional efficiency on some compilers might be gained if we let C be a power of 2, that is, we can let $C = 2^e$ for some positive integer e. If this approach is taken, it appears from empirical testing (not shown) that 2^e for $e \leq 4$ should be avoided, which also implies that C in general should not be too small for given A, B, and order k.

9.1   Design of DW Generators

By the Cayley-Hamilton theorem, the generated vector sequence $(\mathbf{X}_i)_{i \geq 0}$ from this MCG satisfies the recurrence for DW-k

$$\mathbf{X}_i = \alpha_1 \mathbf{X}_{i-1} + \alpha_2 \mathbf{X}_{i-2} + \cdots + \alpha_k \mathbf{X}_{i-k} \bmod p, \quad i \geq k, \tag{9.8}$$

where the multipliers $\alpha_1, \alpha_2, \ldots, \alpha_k$ are defined in Eq. 9.2. Therefore, each of the k sequences taken from each of the k rows in Eq. 9.8 can be viewed as k copies of the same DW-k with different starting seeds. Grothe (1987) was the first to show that MCGs could be used to implement MRGs in this way. Due to the equi-distribution property, every k-tuple of integers in $\mathbb{Z}_p^k$ (except the all-zero tuple) appears exactly once. Hence, the k sequences of DW-k (with different starting seeds) are simply shifts within the same period of DW-k. We can consider this MCG as a parallel implementation of k copies of the same DW-k with some "random" jump-ahead scheme.

Therefore, the recurrence of DW-k can be implemented efficiently and in parallel using the recursion in Eq. 9.6. We recommend generating k numbers at a time and assigning each number to one of k processors. However, there is some strong evidence that the MCG defined by $\mathbf{B}$ in Eq. 9.5 has its own merits as a standalone generator where numbers can be generated one number at a time.

Sharing the Same Characteristic Polynomial

Next, we will prove that the MCG used to implement DW-k has the same corresponding characteristic polynomial $f_{\mathbf{B}}(x)$ as $f(x)$ in Eq. 9.4. Before showing this, we first give the characteristic polynomial $f_{\mathbf{M}}(x)$ of a more general matrix $\mathbf{M}$ of which $\mathbf{B}$ is a special case.

Lemma 9.1 For a given matrix $\mathbf{M}$ of the form

$$\mathbf{M} = \begin{pmatrix} M_1 & 0 & 0 & \ldots & 0 & A \\ M_1 & M_2 & 0 & \ldots & 0 & 0 \\ M_1 & M_2 & M_3 & \ldots & 0 & 0 \\ \vdots & \vdots & \vdots & \ddots & \vdots & \vdots \\ M_1 & M_2 & M_3 & \ldots & M_{k-1} & 0 \\ M_1 & M_2 & M_3 & \ldots & M_{k-1} & M_k \end{pmatrix} \tag{9.9}$$

where A and M_i (for $i = 1, \ldots, k$) are integers, the corresponding characteristic polynomial $f_{\mathbf{M}}(x) = \det(x\mathbf{I} - \mathbf{M})$ is

$$f_{\mathbf{M}}(x) = (x - M_1)(x - M_2) \cdots (x - M_k) - AM_1 x^{k-2}.$$

Theorem 9.1 *Given the matrix* $\mathbf{B}$ *defined in Eq. 9.5, the corresponding characteristic polynomial* $f_{\mathbf{B}}(x) = \det(x\mathbf{I} - \mathbf{B})$ *is*

$$f_{\mathbf{B}}(x) = (x - B)(x - C)^{k-1} - ABx^{k-2}.$$

9.2   Search for DW-k

Now that we have proved that $f_{\mathbf{B}}(x) = f(x)$ in Eq. 9.1, we turn our attention to finding A, B, C such that $f(x)$ is a primitive polynomial over $\mathbb{Z}_p$. Alanen and Knuth (1964) gave necessary and sufficient conditions for determining whether $f(x)$ is primitive or not. The main difficulty is finding the complete factorization of $R(k, p) = (p^k - 1)/(p - 1)$ when k or p is large. There are two common approaches to by-pass the difficulty of the factorization: (a) for a given p one can find k such that $R(k, p)$ is (relatively) easy to factor, usually because $R(k, p)$ has only one huge prime factor and the rest are (relatively) small prime factors or (b) one can consider a prime order k and then find prime p such that $R(k, p)$ is also a prime number.

Applying the first approach, Deng (2005), Deng et al. (2012b) found five k-th degree primitive polynomials for $k \in \{47, 643, 1597, 7499, 20897\}$ and modulus $p = 2^{31} - 1$. This modulus is a popular choice, because the modulus operation can be replaced with more efficient logical operations. Furthermore, it is also the largest (signed) integer that can be stored in a 32-bit computer word. The largest period length for this modulus is approximately $10^{195009.3}$ with equi-distribution up to 20897 dimensions. Following the methods described in these references, for order k, prime modulus $p = 2^{31} - 1$, C in the form of 2^e for $5 \le e \le 9$ (for additional efficiency), and four values of B, we search for A such that DW(k; A, B, C; m) achieves the maximum-period. Table 9.1 tabulates 100 DW(k; A, B, C; $m = 2^{31} - 1$): 5 values of k, 4 values of B, and 5 values of C.

Using the second approach, Deng (2004, 2008), Deng et al. (2012b) found different k ranging from 101 to 25013 and moduli of the form $p = 2^{31} - w$, where w is a positive integer. The largest period length, $p^k - 1$, found so far has reached approximately 10^{233361} with equi-distribution up to 25013 dimensions. Again, following the methods in these references, we search for DW(k; A, B, C; m) with the more flexible form modulus, $p = 2^{31} - w$. For order k, prime modulus p, C in the form of 2^e for $5 \le e \le 9$, and one value of B, we search for A such that DW(k; A, B, C; m) achieves the maximum period. Tables 9.2, 9.3, and 9.4 tabulate 265 DW(k; A, B, C; $m = 2^{31} - w$): 53 values of k and 5 values of C.

Table 9.1 List of A for $DW(k; A, B, C = 2^e; m = 2^{31} - 1)$

k	B	32	64	128	56	512
47	5005	65536	65524	65516	65490	65499
47	10001	65460	65434	65447	65461	65533
47	15090	65455	65519	65472	65505	65468
47	20006	65479	65457	65523	65445	65428
643	5005	65495	63564	64444	63932	62930
643	10001	65326	64001	64562	64797	65510
643	15090	65386	65480	64825	65406	64380
643	20006	63411	65239	64072	63090	64505
1597	5005	64577	65235	64934	65508	63884
1597	10001	63954	63471	64674	63495	65486
1597	15090	63832	65074	63227	65239	64294
1597	20006	65518	65014	64296	61968	65152
7499	5005	64797	59943	63816	61707	61396
7499	10001	58921	49795	52984	55685	60937
7499	15090	56680	53060	64547	64620	53885
7499	20006	62277	59547	58067	59030	50003
20897	5005	21951	59691	49564	33494	62280
20897	10001	27020	53061	35759	29904	37956
20897	15090	34619	36325	58692	49618	31093
20897	20006	47762	44292	57165	13979	62931

9.3 Finding New DW-k via AGM

Utilizing the Automatic Generation Method (AGM) in Deng (2004), once we know the parameters A, B, C for one DW-k, we can quickly find another DW-k of the same order k and modulus p! Recall that, starting with a k-th degree primitive polynomial, AGM quickly finds numerous k-th degree primitive polynomials with the same number of nonzero multipliers as the base primitive polynomial.

To use AGM to find new DW-k, first, choose the desired order k and modulus p and corresponding parameters A, B, C, from Table 9.1 through Table 9.4. A defining feature of DW-k is that its primitive characteristic polynomial $f(x)$ in Eq. 9.1 is equal to $f_\mathbf{B}(x) = \det(x\mathbf{I} - \mathbf{B})$ where $\mathbf{B}$ is defined according to Eq. 9.5. For any nonzero integer z in $\mathbb{Z}_p$, define

Table 9.2 List of A for $DW(k; A, B, C = 2^e; m = 2^{31} - w)$

k	w	B	32	64	128	256	512
101	82845	20000	20028	20066	20093	20050	20217
211	841329	20001	20227	20606	20052	20205	20067
307	52545	20001	20077	20053	20809	20303	20317
401	57189	20000	20184	20266	21862	20106	20131
503	174489	20001	21144	22448	20179	20286	20089
601	1327485	20000	20142	20937	20746	21791	20870
701	220665	20002	20074	21035	20579	20433	20172
809	2010789	20000	20744	21516	20558	20820	20129
907	4400889	20004	22482	20986	20139	20727	20295
1009	2368869	20000	20734	20258	22517	21765	20255
1103	7316361	20002	20637	20813	21623	20535	20201
1201	1113705	20005	21289	22405	21161	20762	20793
1301	1070901	20000	22028	21108	21694	20618	20511
1409	4320189	20000	21545	22719	22407	20942	23501
1511	2771205	20000	21718	20724	22864	21286	21147
1601	368961	20001	20877	21155	22095	22037	20485
1709	1032441	20001	21518	20501	20504	20552	26690
1801	5789241	20002	21325	21222	20049	20892	21539
1901	267321	20002	20336	20779	21070	20050	22125
2003	44961	20001	20499	20606	29363	27302	22033

$$G_{\mathbf{B}}(x) = z^k f_{\mathbf{B}}(z^{-1}x) \bmod p$$
$$= z^k \det(z^{-1}x\mathbf{I} - \mathbf{B}) \bmod p$$
$$= \det(x\mathbf{I} - z\mathbf{B}) \bmod p$$
$$= (x - zB)(x - zC)^{k-1} - (zA)(zB)x^{k-2} \quad \text{(by Theorem 9.1 for } z\mathbf{B}).$$

Letting $A' = zA$, $B' = zB$, and $C' = zC$, then $\alpha_1, \alpha_2, \ldots, \alpha_k$ are given in Eq. 9.2. Specifically, if $\alpha_k = (-1)^{k-1} z^k BC^{k-1} \bmod p$ is a primitive root modulo p, then $G_{\mathbf{B}}(x)$ is the primitive characteristic polynomial of a new DW-k whose multipliers $\alpha_1, \alpha_2, \ldots, \alpha_k$ are defined by A', B', C' (Deng 2004; Theorem 4). The recursion of the new DW-k can be implemented efficiently and in parallel generating numbers k at a time from the recursion in Eq. 9.7. Since we know how many primitive roots exist, then for each modulus p and order k, we know that AGM can produce $\phi(p-1)$ new DW-k from one base DW-k, where $\phi(x)$ denotes the Euler "totient" function, which gives a number of integers between 1 and x that are relatively prime to x.

Table 9.3 List of A for $DW(k; A, B, C = 2^e; m = 2^{31} - w)$

k	w	B	32	64	128	256	512
2111	3536385	20005	20326	21705	23674	20213	23241
2203	6043089	20002	23370	24475	25871	24984	21072
2309	340185	20003	22094	20623	20372	26437	22412
2411	9256449	20006	20975	21355	22754	25674	20429
2503	13539249	20001	21325	20477	22106	21521	25097
2609	8811681	20001	20761	22499	23285	21895	20796
2707	1113585	20004	20521	20016	20403	20994	24184
2801	1095609	20002	20363	25844	20090	20993	25253
2903	14055825	20003	20159	20246	23773	20718	23791
3001	3058401	20001	28132	20617	21599	21230	23706
3109	6741129	20002	20692	22733	23316	21300	25581
3203	4718889	20001	22833	25436	38018	21850	21251
3301	14881185	20002	28598	26152	20943	22581	20220
3407	6243009	20001	20904	29014	31949	22005	20327
3511	1412961	20001	21747	25713	20470	26032	23604
3607	1026585	20003	20773	22329	20577	28986	29789
3701	11576625	20002	21346	22672	29143	27596	21854
3803	32058129	20002	21522	21082	20997	21753	27771
3907	17381649	20001	23800	20442	24531	24191	20518
4001	4412481	20001	24549	20465	24707	22823	20072

One advantage of applying AGM to DW-k is that, for a given order k and modulus p, it gives users the liberty to find a new DW-k if for some reason they need more parameters in addition to those in Table 9.1 through Table 9.4.

9.3.1 Evaluating the Spectral Test and Its Limitations

Often researchers want one measure to compare a set of random number generators to each other. For years, the spectral test was this theoretical measure as "not only do all good generators pass this test, all generators are now known to be bad actually *fail* it" Knuth (1998). For a description of the spectral test, see Kao and Tang (1997), Knuth (1998), L'Ecuyer (1990), L'Ecuyer and Couture (1997). However, with the expansion of large-order, equi-distributional pseudo random number generators in the literature, this theoretical measure now seems less "golden". In MRGs, for a given order k and dimension $t > k$, the spectral test measures the greatest distance between two adjacent $(t - 1)$-dimensional hyperplanes

Table 9.4 List of A for $DW(k; A, B, C = 2^e; m = 2^{31} - w)$

k	w	B	32	64	128	256	512
5003	1259289	20001	27764	20766	20859	24425	21082
6007	9984705	20001	20113	24411	30649	21365	26515
7001	610089	20003	23538	30769	31220	20721	24699
8009	5156745	20001	24596	30764	29030	28750	26236
9001	7236249	20002	34768	24425	22364	32772	24052
10007	431745	20003	20258	63815	33089	47201	28169
11003	1276425	20001	45421	51250	25568	21659	23189
12007	37532781	20000	37739	27242	30994	34999	22036
13001	71128005	20003	52978	21790	24522	38227	28803
14009	626301	20000	63410	26334	29856	77138	27117
15013	8996265	20003	72834	22198	38433	37498	20490
20011	26131941	20000	43502	37043	31430	69099	24899
25013	11538909	20000	25118	24538	57529	87512	66488

in a family of parallel hyperplanes that cover all t-tuples of successive, overlapping output from the entire period of the MRG. These t-tuples form a t-dimensional lattice hypercube.

When this test is applied to the first-order linear recurrence of LCGs, we see that the 2-dimensional plot of successive, overlapping pairs does not appear as random looking as we would expect. A similar structure is found when we take successive, overlapping triplets and plot them in 3-dimensions. In fact, we know the structure will be bad for all dimensions $t \geq 2$. When considering large-order MRGs with orders k in the 100 s, 1000 s, or 10000 s, it is more difficult to understand the practical effect of this test for dimensions greater than the order, especially when we consider $k = 20897$ or $k = 25013$.

Furthermore, if the simulation scientist knew she needed equi-distribution up to order t, she can often just choose a larger order for her favorite generator with the equi-distribution property. For MRGs, she would only run into problems if she needed equi-distribution beyond order 25013. Equi-distribution generators for larger orders will necessarily be "better" than those for smaller orders. And at the moment, this is the best we can say as there is no agreed upon figure of merit, for large-order MRGs.

The spectral test can also be applied not only to successive indexes of the generated output, it can also be applied to a "jump-around" index, called a lacunary index L'Ecuyer (1997). To find a particular lacunary index, such that the spectral test performs "poorly", one usually inspects the structure of the given generator and looks for a way to exploit it. Once a "poor" lacunary index is known, the generator will very likely perform poorly when generating output according to this index, which will resemble some pattern similar to use some numbers, skip some, use some, skip some, and so on. As long as the generator does not generate numbers according to this pattern, it will perform well.

According to L'Ecuyer (1997), a necessary (but not sufficient) condition for a "good" MRG is that the sum of the squares of coefficients, $\sum_{i=1}^{k} \alpha_i^2$, should be large. Given that DW-k has k nonzero multipliers, then implementing the k copies of DW-k is likely to have excellent spectral test values when using a successive index. Furthermore, since there is no simple relationship among these k nonzero multipliers and k copies of DW-k are implemented in parallel, it appears very difficult to find a lacunary index such that the k copies perform poorly on the spectral test.

MRGs with Many Non-Zero Terms and Efficient Implementation

10

A good pseudo-random number generator (RNG) of today should have a sufficiently long period, fast generating speed, and high-dimensional uniformity—namely, for some high dimension t, all t-tuples of successively generated numbers are distributed uniformly in the t-dimensional space. Good RNGs should generate streams of numbers that "feel" and "look" random; therefore, they should also have satisfactory empirical performance. In addition to these requirements, the strong demand of parallel processing in today's computing environment requires a mechanism to automatically produce good, parallel RNGs or alternatively enable a good RNG to generate "independent" streams of pseudo-random numbers for multiple processors in parallel.

For several decades, a large amount of research efforts in random number generation have been devoted to searching for the "best" RNGs. Early on, the linear congruential generator (LCG) proposed by Lehmer (1951) was established as the RNG of choice. This first-order linear recurrence method is indeed one of the most efficient RNGs to date. The search for the "best" RNG (inevitably) led to the so-called "minimal standard" LCG published in a widely cited paper, "*Random Number Generators: Good Ones are Hard to Find*", written by Park and Miller (1988). However, the LCG has a short period (by today's standards), poor empirical performance, and inadequate uniformity in dimensions higher than one. These and other poor properties have earned the LCG a reputation as an unsuitable generator for modern simulation tasks. In "*Good random number generators are (not so) easy to find*", Hellekalek (1998) made the point that good RNGs exist, but you need to know what to look for; furthermore, good RNGs are definitely not LCGs.

For a long time, the multiple recursive generator (MRG) has been considered by many to be the great contender for replacing the established LCG. An MRG of order k generates pseudo-random numbers sequentially with a k-th order linear recurrence under a prime

L. Deng et al., *Random Number Generators for Computer Simulation and Cyber Security*, Synthesis Lectures on Mathematics & Statistics, https://doi.org/10.1007/978-3-031-76722-7_10

modulus p. Maximum-period MRGs have two very desirable properties: a huge period of $p^k - 1$ and the nice equi-distribution property up to order k. They also have a strong statistical justification and excellent empirical performance. Indeed, concerning the output of these MRGs, Knuth (1998) said "*all known evidence indicates that the result will be a very satisfactory source of random numbers.*"

However, finding a maximum-period MRG can be very time-consuming when the order k and/or prime modulus p are large. Furthermore, even for a moderate order k, implementing maximum-period MRGs with many nonzero coefficients efficiently or in parallel is not trivial. Most research in this area has tried to sidestep these difficulties by searching for MRGs with many nonzero terms but of certain special structures such that the linear recurrence can be implemented with a higher-order MRG of few nonzero terms—fewer nonzero terms require fewer costly multiplications when implementing the recursion (see, e.g., Deng 2008). However, when using particular non-successive sequences, there are concerns regarding the empirical and theoretical performances of MRGs with few nonzero terms. It is generally perceived that maximum-period MRGs with many nonzero terms (and no special structure) would be free of this imperfection. In other words, a maximum-period MRG with many nonzero terms would have all the qualities of a good RNG except for an efficient implementation and a parallelization mechanism.

The contributions are as follows:

- Using well-known theorems from number theory and algebra we show how to find, from one single given maximum-period MRG, all of the maximum period MRG's of the same order k and modulus p (this set is denoted MRG(k, p)). This new method, which we call algorithm AGM-2, is presented in Sect. 10.2.1. As an application of this method, we show how starting with an instance of the DX-k-2 generator, we can find generators with much better performance on the spectral test. This is presented in Sect. 10.2.2.
- A novel method for efficient implementation of a given MRG in MRG(k, p) is presented. The method is through a specific MCG which will produce "k parallel copies" of MRG streams for k processors, each with different random starting seeds. In particular, the number of operations to generate each new pseudo-random number on a processor is reduced from k to 2 (or less). This is given in Sect. 10.3.3. Furthermore, in Sect. 10.3.4 we show how to choose certain coefficients involved so that the method can be implemented efficiently on a computer using bit shifting.

This chapter is organized as follows. In Sect. 10.1.1, we present some background material on maximum-period MRGs, including their good qualities and certain difficulties in finding them. In Sect. 10.2, we describe automatic generation methods. These methods start with a maximum-period MRG, and produce new generators that are also maximum-period MRGs. We present a known automatic generation method from Deng (2004) and a new one that is even more general. Furthermore, we find several MRGs with many nonzero terms for various orders k and the prime modulus $p = 2^{31} - 1$ to demonstrate that those of the same order k tend to have a similar (theoretical) performance on the spectral test. In Sect. 10.3, we give an efficient and parallel implementation for any general maximum-period MRG with many nonzero terms. Finally, we discuss the implications along with some other considerations.

10.1 Background

10.1.1 Multiple Recursive Generators (MRG)

An MRG is defined by a k-th order linear recurrence as

$$X_i = \alpha_1 X_{i-1} + \alpha_2 X_{i-2} + \cdots + \alpha_k X_{i-k} \quad \text{mod } p, \ i \geq k, \tag{10.1}$$

where $\alpha_1, \alpha_2, \ldots, \alpha_k$ are integers in $\mathbb{Z}_p$, $\alpha_k \neq 0$, and any k values, $(X_0, X_1, \ldots, X_{k-1}) \neq (0, 0, \ldots, 0)$, can be used as the starting seeds. It is well known that the MRG in Eq. 10.1 has the maximum period, $p^k - 1$, if and only if its characteristic polynomial

$$f(x) = x^k - \alpha_1 x^{k-1} - \alpha_2 x^{k-2} - \cdots - \alpha_k \tag{10.2}$$

is a k-th degree primitive polynomial over $\mathbb{Z}_p$.

For the MRG defined in Eq. 10.1, we can define its companion matrix as

$$\mathbf{M}_f = \begin{pmatrix} 0 & 1 & 0 & \ldots & 0 \\ 0 & 0 & 1 & \ldots & 0 \\ \vdots & \vdots & \vdots & \ddots & \vdots \\ 0 & 0 & 0 & \ldots & 1 \\ \alpha_k & \alpha_{k-1} & \alpha_{k-2} & \ldots & \alpha_1 \end{pmatrix}. \tag{10.3}$$

It is straightforward to see

$$f(x) = \det(x\mathbf{I} - \mathbf{M}_f) \quad \text{mod } p. \tag{10.4}$$

For the remainder of this chapter, we let MRG(k, p) denote the class of maximum-period k-th order MRGs with prime modulus p. There are exactly $\phi(p^k - 1)/k$ primitive polynomials of degree k (see, e.g., Knuth 1998), where $\phi(x)$ is the Euler totient function, giving the number of integers between 1 and x that are relatively prime to x. It is well known that if the complete prime factorization of x is $p_1^{r_1} p_2^{r_2} \cdots p_c^{r_c}$, then

$$\phi(x) = x \prod_{i=1}^{c} \left(1 - \frac{1}{p_i}\right).$$

See, for example, Knuth (1998). Therefore, we can see that the total number of primitive polynomials of degree k grows exponentially in k because it is (roughly) of the same order as $O((p^k - 1)/k)$.

It is widely known that generators in MRG(k, p) have strong statistical justifications. Deng and George (1990), Deng et al. (1997) and excellent empirical performance Deng (2005), Deng et al. (2012a), L'Ecuyer and Simard (2007). Furthermore, all maximum-period MRGs have an extremely long period of $p^k - 1$ and the equi-distribution property up to order k; that is, during its entire period, every t-tuple ($1 \leq t \leq k$) of integers in $\mathbb{Z}_p^t$ appears exactly the same number of times (p^{k-t}), with the exception of the all-zero tuple that appears once less (see, for example, Lidl and Niederreiter 1994, Theorem 7.43). The equi-distribution property implies that t-dimensional subsequences obtained from the successive output of a generator in MRG(k, p) are good approximations to a uniform distribution in the t-dimensional space for $t \leq k$. For a fixed modulus p, the large period and equi-distribution properties become more and more advantageous as order k increases.

10.1.2 A Class of Efficient MRG(k, p)

When the k-dimensional state vector is close to the zero vector, the subsequent numbers generated may stay within a neighborhood of zero for quite a while before they can break away from this "near-zero land", a property apparently not desirable in the sense of randomness. Consequently, two generated sequences using the same DX generator with nearly identical state vectors may not depart from each other quickly enough. This "bad initialization effect" was first observed by Panneton et al. (2006) for MT19937. In addition, as pointed out in L'Ecuyer and Simard (2014), L'Ecuyer et al. (2021, 2017), their theoretical property over a higher (larger than k) dimensional space tends to be poor when compared to a general MRG with many nonzero terms.

To overcome this "bad initialization effect", one can design classes of MRGs with many nonzero terms. In general, the direct implementation of this type of generator usually is inefficient; yet it is still possible to have an efficient implementation for certain classes of MRGs (see, e.g. Deng and Shiau 2015; Deng et al. 2023). To achieve efficiency, these generators were designed in a way that they can be implemented via a recursive equation that is one order higher and contains only a few nonzero terms. However, the efficient design as such still suffers from the same problems that an MRG with few nonzero coefficients has but over the spaces of dimension larger than $k + 1$ (instead of k).

10.1.3 Spectral Test for Comparing MRGs

As aforementioned, all generators in MRG(k, p) have the nice equi-distribution property in t-dimensional space up to $t \leq k$. The spectral test is a theoretical test that provides a measure of uniformity in dimensions beyond the order k of the MRG. It is often used to rank MRGs of the same order. Consider all the successive sequences of length t generated from the recurrence Eq. 10.1, $S_n = (X_n, X_{n+1}, \ldots, X_{n+t-1})$ for $n = 0, 1, \ldots, \rho - 1$, where $\rho \equiv p^k - 1$ is the period length. The spectral test computes the largest distance between adjacent parallel hyperplanes among families of parallel hyperplanes that cover all the points $\{S_n\}$ in the t-dimensional space with $t > k$ (see, e.g., Knuth 1998; L'Ecuyer 1997). This largest distance is referred to as the *spectral distance* in the literature and is often denoted by $d_t(k)$.

When $t > k$, the spectral distance is a measure of the uniform spread of the t-tuples across the t-dimensional space; and a small spectral distance implies a more uniform coverage. Therefore, a large $d_t(k)$ is considered "bad," for it would take only a relatively small number of parallel $(t - 1)$-dimensional hyperplanes to cover all the t-dimensional lattice points produced by the generator; and, consequently, the generator is often said to have a bad lattice structure in dimension t because of the large distance between hyperplanes. Clearly, when t is much larger than k, the spectral distance $d_t(k)$ becomes so large that no MRGs (of fixed order k) can be considered "good" when evaluated by $d_t(k)$.

To compute the spectral test of a general MRG, we can use the similar procedures proposed in Kao and Tang (1997), Knuth (1998), L'Ecuyer (1997), L'Ecuyer and Couture (1997).

10.2   Automatic Generation Methods

For a large k, it is very time-consuming to check if a given k-th degree polynomial $f(x)$, as given in Eq. 10.2, is a primitive polynomial. However, when given a known primitive polynomial $f(x)$, it is quite easy to produce a lot of k-th degree primitive polynomials.

We first describe an automatic generation method (AGM) presented in Deng (2004); we call this AGM-1. Then, we describe a new method (AGM-2) which is more general since it works for any prime p, while AGM-1 requires p to satisfy a condition, see below. Then, we consider the MRG-k generators that result from AGM-2 applied to the well-known DX-k-2 generators.

AGM-1 is based on the theoretical results of Deng (2004), which showed that, under the condition that $R(k, p) = (p^k - 1)/(p - 1)$ be a prime number and $f(x)$ in Eq. 10.2 be a k-th degree primitive polynomial, if $c^{-k}\alpha_k$ is a primitive element modulo p, then

$$G(x) = c^{-k} f(cx) \quad \text{mod } p \tag{10.5}$$

is a k-th degree primitive polynomial.

It is straightforward to find the explicit formula for $G(x)$ as

$$G(x) = x^k - G_1 x^{k-1} - G_2 x^{k-2} - \cdots - G_k,$$

where

$$G_j = c^{-j}\alpha_j \quad \mathrm{mod}\ p \text{ for } j = 1, 2, \ldots, k.$$

Note that $G(x)$ is a primitive polynomial that has the same number of nonzero terms as $f(x)$. Therefore, if we choose the $f(x)$ corresponding to an MRG with two nonzero terms, the MRG corresponding to each generated primitive polynomial $G(x)$ is also an efficient two-term generator.

To avoid checking whether $c^{-k}\alpha_k$ is a primitive element modulo p, Deng (2004) further proposed the following "automatic generation" procedure based on a similar "leap-frog procedure" proposed in Deng et al. (1994) for the LCG:

Let R be a primitive element modulo $p - 1$ with $\gcd(R, p - 1) = 1$. The following procedure will randomly generate a sequence of maximum-period MRGs for parallel processing.

Algorithm 10.1: Algorithm AGM-1

1. Whenever a new processor is initiated, generate an r_n by the following equation:

$$r_n = R\,r_{n-1} \bmod (p - 1), n \geq 1 \text{ with } r_0 = 1.$$

2. Calculate d_n and then c_n for the new processor by

$$d_n = k^{-1}(r_n + 1) \bmod (p - 1) \text{ and } c_n = B^{d_n} \bmod p.$$

3. With the given c_n, compute the primitive polynomial $G(x)$ by

$$G(x) = c_n^{-k} f(c_n x) \bmod p;$$

   equivalently, compute its coefficients by

$$G_j = c_n^{-j}\alpha_j \quad \mathrm{mod}\ p \text{ for } j = 1, 2, \ldots, k.$$

4. The new processor can use the newly constructed maximum-period MRG corresponding to the characteristic polynomial $G(x)$ as

$$X_i = G_1 X_{i-1} + \cdots + G_k X_{i-k} \bmod p.$$

Remarks

1. Algorithm AGM-1 requires $(p^k - 1)/(p - 1)$ to be a prime number. Therefore, it is not applicable for MRGs with $p = 2^{31} - 1$, since $(p^k - 1)/(p - 1)$ is usually not a prime number.
2. If $f(x)$ is a primitive polynomial corresponding to efficient MRGs of few nonzero terms like DX-k-s generators, then the MRGs corresponding to the generated $G(x)$'s are also

efficient because they have the same number of nonzero terms. However, as explained in Kao and Tang (1997), generators in MRG(k, p) with few nonzero terms generally perform worse than those with many nonzero terms over high-dimensional spaces. In addition, large-order MRGs with few nonzero term tend to be too slow to "escape from zero states" (see also L'Ecuyer and Simard 2014).

3. This procedure is useful to generate many $G(x)$'s from a single primitive polynomial $f(x)$ with many nonzero terms. However, the straightforward implementation for the corresponding MRG can be quite inefficient.

Next, we describe another automatic generation method that will work for any prime number p.

10.2.1 A New Automatic Generation Method (Algorithm AGM-2)

Let $f(x)$ in Eq. 10.2 be a k-th degree primitive polynomial with companion matrix $\mathbf{M}_f$ as in Eq. 10.3. Define

$$f_r(x) = \det(x\mathbf{I} - \mathbf{M}_f^r) \quad \mod \ p. \tag{10.6}$$

The second automatic generation algorithm is based on some well-known results in the theory of finite field (see, e.g., Zierler 1959; Golomb 1982). Specifically, it is known that $f_r(x)$ is a primitive polynomial if and only if r is relatively prime to $p^k - 1$. In addition, $f_r(x)$ is equal to $f_{r'}(x)$ only when $r = r'p^s \mod (p^k - 1)$ for some integer $s = 0, 1, \ldots, k - 1$. This implies that there are k duplications for each r; hence there exist $\phi(p^k - 1)/k$ k-th degree primitive polynomials. Once a primitive polynomial $f(x)$ is available, a new k-th degree primitive polynomial corresponding to an r coprime to $p^k - 1$, $f_r(x)$, can be obtained by raising the companion matrix $\mathbf{M}_f$ of $f(x)$ to the power of r (see, e.g., Haramoto et al. 2008a, b) and then finding $\det(x\mathbf{I} - \mathbf{M}_f^r) \mod p$.

While it is possible to find all k-th degree primitive polynomials with just one primitive polynomial $f(x)$ in theory, it is practically infeasible to find all of them. In practice, we can restrict the search of r in a range as $[a, b]$ with small a and b ($1 < a < b < p$). Since any two values r and r' in such a small range will not satisfy $r = r'p^s \mod (p^k - 1)$, duplicate primitive polynomials (thus duplicate MRGs) will not be produced. Moreover, small values of r and the corresponding $f_r(x)$ will be easier to find/compute in practice.

The following procedure will generate a sequence of maximum-period MRGs from a given primitive polynomial $f(x)$.

Algorithm 10.2: The Second Automatic Generation Method (AGM-2)

---

1. Find a sequence of numbers in $[a, b]$ (say, $r_1, r_2, \ldots$) that are relatively prime to $p^k - 1$.
2. For each r_j, compute $G(x) = f_{r_j}(x) = \det(x\mathbf{I} - \mathbf{M}_f^{r_j}) \mod p$ as in Eq. 10.6, and then construct the maximum-period MRG corresponding to the characteristic polynomial $G(x)$ as

$$X_i = G_1 X_{i-1} + \cdots + G_k X_{i-k} \mod p.$$

Remarks

1. Algorithm AGM-2 may not be suitable when the order k is large, because it would be time-consuming to compute $f_r(x)$, which requires raising a large $k \times k$ matrix, $\mathbf{M}_f$, to some power r, then finding the determinant of a large $k \times k$ matrix, namely, $f_r(x) = \det(x\mathbf{I} - \mathbf{M}_f^r)$.
2. Algorithm AGM-2 is particularly useful for $p = 2^{31} - 1$ and small order k. We choose $k \leq 24$ because the complete factorization of $(p^k - 1)/(p - 1)$ had been found for these k. With that, we find some DX-k-2 generators to serve as the starting MRGs.
3. Although each starting MRG has only two nonzero terms, the MRGs corresponding to $f_r(x)$ for $r \neq 1$ have many nonzero terms and we will show in the next subsection that all of them have a better spectral test value than that of their starting DX generators. This confirms the general argument that generators in MRG(k, p) with many nonzero terms perform better on the spectral test than those with few nonzero terms.
4. The feature of many nonzero terms for the MRGs generated by Algorithm AGM-2, on the other hand, slows down the generating speed. To circumvent this drawback, we develop a novel efficient implementation for such MRGs. Therefore, contingent on an efficient and parallel implementation, a profusion of good RNGs can be constructed.

10.2.2 Spectral Test for MRG-k Produced by Algorithm AGM-2

For the remainder of this section, we assume that the prime modulus used is $p = 2^{31} - 1$, the most popular modulus for 32-bit generators. In this experiment, we consider $k = 5, 6, \ldots, 24$. For each k, we find a generator in DX-k-2 to be the starting generator. To find some r's with $\gcd(r, p^k - 1) = 1$, we choose a small range for r with $a = 100$ and $b = 200$, in which there are slightly over 20 such r's for each k. Let $f(x)$ denote the primitive polynomial of the starting DX-k-2 generator, compute $f_r(x) = \det(x\mathbf{I} - \mathbf{M}_f^r) \mod p$ for the r's in $[a, b]$ with $\gcd(r, p^k - 1) = 1$ to produce MRG-k generators via Algorithm AGM-2.

To evaluate the MRG-k generators produced, we calculate $d_t(k)$ for dimensions $t = k + 1$ and $t = k + 10$ for each of the MRG-k generators using the methods described in Kao and

Table 10.1 Spectral test of the starting DX-k-2 versus the produced MRG-k's for various order k

	DX-k-2		MRG-k			
k	B	$d_{k+1}(k)$	min $d_{k+1}(k)$	max $d_{k+1}(k)$	min $d_{k+10}(k)$	max $d_{k+10}(k)$
24	1011139	2.0169e-05	5.4681e-10	1.1147e-09	4.2325e-08	1.1493e-07
23	1010738	8.5545e-05	6.5150e-10	1.1357e-09	7.9450e-08	1.5698e-07
22	1010598	1.9042e-05	5.4479e-10	1.3708e-09	9.8564e-08	1.9183e-07
21	1010675	6.8704e-05	7.0677e-10	1.1904e-09	1.3134e-07	2.6068e-07
20	1011093	3.6123e-05	7.0100e-10	1.4115e-09	1.4896e-07	3.3066e-07
19	1011385	2.1093e-05	7.8290e-10	1.4744e-09	2.5707e-07	4.7395e-07
18	1010715	2.9729e-05	8.8826e-10	1.5527e-09	3.0252e-07	6.4520e-07
17	1010547	2.8752e-05	9.2199e-10	2.1985e-09	4.6684e-07	9.0989e-07
16	1011236	3.1188e-05	1.0428e-09	2.1323e-09	6.8931e-07	1.2099e-06
15	1010677	4.3985e-05	1.2033e-09	2.0694e-09	9.1564e-07	1.7059e-06
14	1010568	2.6494e-05	1.3987e-09	2.9181e-09	1.6032e-06	2.7158e-06
13	1010667	4.2776e-05	1.6456e-09	2.8490e-09	2.4466e-06	4.6174e-06
12	1010866	9.3999e-05	1.7686e-09	3.2046e-09	4.1708e-06	6.2978e-06
11	1010397	2.6414e-05	2.3066e-09	3.4175e-09	6.5306e-06	1.4636e-05
10	1010571	3.4786e-05	3.0760e-09	5.8136e-09	1.1926e-05	1.9623e-05
9	1010274	2.0613e-05	3.6131e-09	8.2099e-09	2.3410e-05	3.4665e-05
8	1010661	3.9999e-05	4.5075e-09	8.1566e-09	4.5410e-05	9.0088e-05
7	1010167	1.9872e-05	6.1129e-09	1.2974e-08	8.7102e-05	0.00017899
6	1010233	1.7798e-05	1.0093e-08	2.0482e-08	0.00022300	0.00033965
5	1010152	3.3185e-05	1.6960e-08	3.7570e-08	0.00049972	0.00087323

Tang (1997), L'Ecuyer (1997), L'Ecuyer and Couture (1997), Knuth (1998); furthermore, we compare their $d_t(k)$ against their starting DX-k-2 generators.

Table 10.1 tabulates the ranges of $d_{k+1}(k)$ and $d_{k+10}(k)$ for the MRG-k generators produced. The multiplier and $d_{k+1}(k)$ of each DX-k-2 are also included. We further use boxplots to depict for each order k the distribution of $d_{k+1}(k)$ and $d_{k+10}(k)$ of the produced MRG-k's in Figs. 10.1 and 10.2, respectively, on a $\log_{10}$ scale.

From Table 10.1, Figs. 10.1, and 10.2, it is clear to see that the values of the maximum gap between adjacent hyper-planes (in dimensions $k+1$ and $k+10$) for the vast majority of the MRG-k generators are reasonably small. In contrast, the maximum gap for the DX-k-2 stays in the order of 10^{-5} for various values of k. This means that we can use Algorithm AGM-2 to produce MRGs with a much better spectral test property than DX generators (and other MRGs with few nonzero terms). For example, the maximum gap over dimension $t = k + 1$ for the DX-k-2 generator is larger than those of the MRG-k generators by a factor of $O(10^3)$ to $O(10^5)$ folds. This ratio clearly gets more profound as k gets larger. It is interesting to

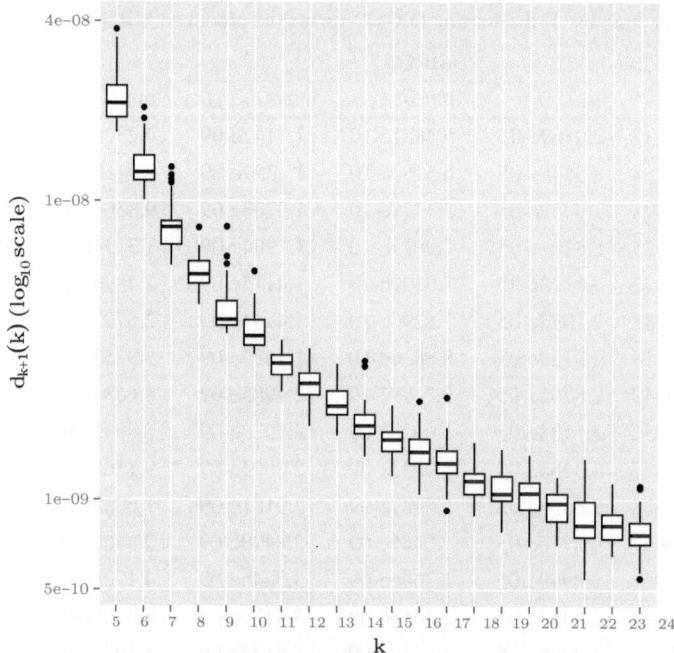

Fig. 10.1 Boxplots of the $d_{k+1}(k)$ values of the produced MRG-k generators for $k = 5, 6, \ldots, 24$

observe that, for each k, the ratio of maximum and minimum $d_t(k)$ is around 2 for both $d_{k+1}(k)$ and $d_{k+10}(k)$, implying similar performance on spectral test among the produced MRG-k generators. This may lend some support to the claim that it is unnecessary to choose a particular "best" MRG-k.

From Table 10.1, Figs. 10.1, and 10.2, we see a general trend: MRG-k generators for larger k tend to have smaller $d_t(k)$ values for $t = k + 1$ and $k + 10$, especially in the $\log_{10}$ scaling. Although not shown here, this trend persists across $d_{k+2}(k), \ldots, d_{k+9}(k)$. Also, as expected, it is observed that the maximum gap $d_{k+10}(k)$ is larger than $d_{k+1}(k)$ for all k.

Next, we address the only remaining problem by providing an efficient and parallel implementation for our proposed MRGs with many nonzero terms.

10.3   Efficient Implementation of MRGs with Many Non-Zero Terms

For the remainder of this chapter, without introducing new notation, we will let $f(x)$ as in Eq. 10.2 be a primitive polynomial with the coefficients α_i, $1 \le i \le k$, all nonzero. As shown earlier, the MRG corresponding to such an $f(x)$ tends to have better spectral test values but its direct/straightforward implementation can be slow. In this section, we show that a suitable choice of the matrix congruential generator (MCG), which is a k-dimensional

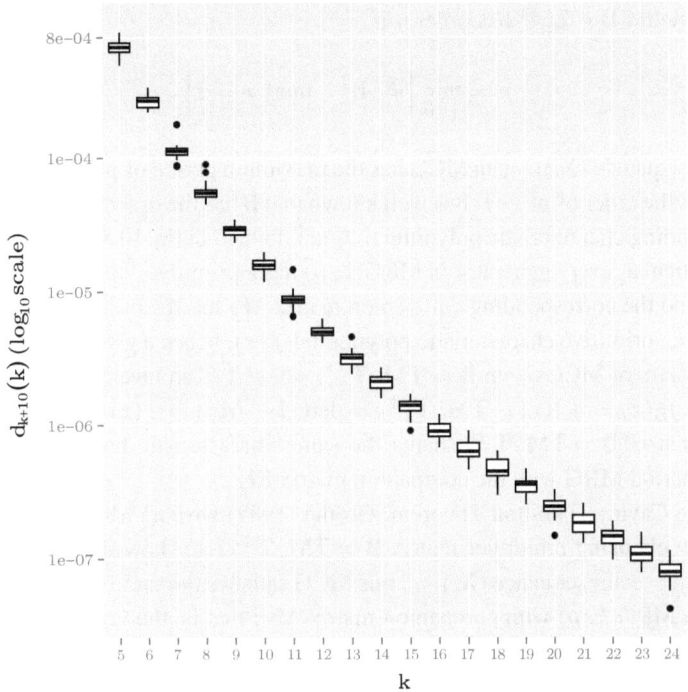

Fig. 10.2 Boxplots of the $d_{k+10}(k)$ values of the produced MRG-k generators for $k = 5, 6, \ldots, 24$

generalization of the LCG by embedding the k multipliers in a $k \times k$ matrix, can be used to implement any generator in MRG(k, p) in parallel as well as to enhance the generating efficiency of the MRG.

10.3.1 MCG Implementation of an MRG

The MCG is a natural k-dimensional extension of the LCG as defined by

$$\mathbf{X}_i = \mathbf{B}\mathbf{X}_{i-1} \quad \mathrm{mod}\ p,\ \ i \geq 1, \tag{10.7}$$

where $\mathbf{X}_i$ is a k-dimensional vector in $\mathbb{Z}_p^k$ and the multiplier matrix $\mathbf{B}$ is a $k \times k$ matrix in $\mathbb{Z}_p^{k \times k}$. The MCG was considered by many researchers in the literature, for example, Franklin (1964), Grothe (1987), L'Ecuyer (1990), Niederreiter (1986). The characteristic polynomial of an MCG is defined as

$$f_{\mathbf{B}}(x) = \det(x\mathbf{I} - \mathbf{B}) \quad \mathrm{mod}\ p. \tag{10.8}$$

An integer matrix $\mathbf{B} \in \mathbb{Z}_p^{k \times k}$ has order n if

$$n = \min_{j>0} \left\{ j : \mathbf{B}^j \quad \mathrm{mod} \ p = \mathbf{I} \right\}.$$

As a vector sequence $(\mathbf{X}_i)_{i \geq 0}$, an MCG has the maximum period of $p^k - 1$ *if and only if* the matrix $\mathbf{B}$ has the order of $p^k - 1$. It is well known that $\mathbf{B}$ has the order of $p^k - 1$ *if and only if* its corresponding characteristic polynomial $f_\mathbf{B}(x)$ defined in Eq. 10.8 is a primitive polynomial. In particular, every generator in MRG(k, p) has a primitive characteristic polynomial, say, $f(x)$, and the corresponding companion matrix $\mathbf{M}_f$ has the order of $p^k - 1$.

For a given primitive characteristic polynomial $f(x)$, hence a given $\mathbf{M}_f$, Grothe (1987) proposed a class of MCGs with $\mathbf{B} = \mathbf{TM}_f \mathbf{T}^{-1}$, where $\mathbf{T}$ is an invertible $k \times k$ matrix over $\mathbb{Z}_p^{k \times k}$. Since $f_\mathbf{B}(x) = \det(x\mathbf{I} - \mathbf{TM}_f \mathbf{T}^{-1}) = \det(x\mathbf{I} - \mathbf{M}_f) = f(x)$ over $\mathbb{Z}_p$, an MCG with multiplier matrix $\mathbf{B} = \mathbf{TM}_f \mathbf{T}^{-1}$ shares the same characteristic polynomial as that of a maximum-period MRG with the companion matrix $\mathbf{M}_f$.

Using the Cayley-Hamilton Theorem, Grothe (1987) gave an MCG implementation of the MRG by choosing multiplier matrix $\mathbf{B} = \mathbf{TM}_f \mathbf{T}^{-1}$. He showed that when taken as a k-dimensional vector sequence $(\mathbf{X}_i)_{i \geq 0}$, this MCG satisfies the same recursion as that of the generator in MRG(k, p) with companion matrix $\mathbf{M}_f$; that is, the vector sequence $(\mathbf{X}_i)_{i \geq 0}$ satisfies the following recursion:

$$\mathbf{X}_i = \alpha_1 \mathbf{X}_{i-1} + \alpha_2 \mathbf{X}_{i-2} + \cdots + \alpha_k \mathbf{X}_{i-k} \quad \mathrm{mod} \ p, \ i \geq k. \tag{10.9}$$

Therefore, the k sequences taken from each of the k rows in Eq. 10.9 can be viewed as k copies of the same MRG with different starting seeds. More detailed theoretical relations between MCGs and MRGs is considered in Sect. 10.4.

10.3.2 Two Ways to Output MCG Variates

There are two natural ways to output variates from the vector variates generated by the MCG—column-wise and row-wise.

The column-wise way produces a sequence of k-dim vectors $\mathbf{X}_i$, $i = 1, 2, \ldots$. For a maximum-period MCG, this vector sequence will repeat after the period of $p^k - 1$ is reached. Therefore, any nonzero k-dimensional vector in $\mathbb{Z}_p^k$ will appear only once and this uniformity property over the k-dimensional cube of $\mathbb{Z}_p^k$ is a requirement for $\mathbf{X}_i/p$ to resemble a k-dimensional random vector uniformly distributed over $[0, 1]^k$. Additionally, choosing any t-dimensional sub-vector $(t \leq k)$ from $\mathbf{X}_i$ for $i = 1, 2, \ldots$ gives a set of t-dimensional points that has the equi-distributional property over t-dimensional space, similar to those generated by a maximum-period MRG. That is, all t-dimensional nonzero sub-vectors will occur $p^{(k-t)}$ times and all t-dimensional zero sub-vectors will occur $p^{(k-t)} - 1$ times. This "column-wise" output method is particularly suitable for some modern vector processors or software packages such as R.

The row-wise method produces k output streams corresponding to the k-rows in the generating Eq. 10.9. As mentioned earlier, these streams can be viewed as outputs of the same MRG with different starting shifts. Since the period length ($p^k - 1$) is huge, a set of random starting seed vectors can produce reasonably large shifts among k different streams. This "row-wise" output method is particularly suitable in producing streams in parallel for multiple processors.

Next, we consider the key issue of choosing a particular matrix multiplier $\mathbf{B}$ such that the corresponding MCG is efficient to implement in either "column-wise" or "row-wise" way for outputting variates.

10.3.3 Efficient and Parallel MCG Implementation of MRG(k, p)

Due to the equi-distribution property, every k-tuple of integers in $\mathbb{Z}_p^k$ except for the all-zero tuple appears exactly once. Hence, the aforementioned k MRG sequences (with different starting seeds) are simply shifts within the same MRG cycle. Clearly, different choices of matrix multiplier $\mathbf{B}$ will result in different shifts. Typically, due to the huge period length of MRG(k, p), a "random" choice of $\mathbf{B}$ (or $\mathbf{T}$) tends to yield k sequences that are far apart in the MRG cycle. It can be shown that, on average, the k sequences are at least $O((p^k - 1)/k^2)$ numbers apart. With this in mind, we can consider an MCG as a parallel implementation of k copies of the same generator in MRG(k, p) with some "random" jump-ahead scheme.

From Grothe's parallel MCG implementation, it is not clear how to choose a matrix $\mathbf{T}$ in order to have an efficient generating recursion for an MCG. The matrix $\mathbf{B} = \mathbf{T}\mathbf{M}_f\mathbf{T}^{-1}$ generally has $O(k^2)$ nonzero terms and hence requires $O(k^2)$ multiplications to produce a single k-dimensional vector. Thus, on average, the number of multiplications needed per variate is of order $O(k)$. This is clearly not efficient.

If we choose $\mathbf{T}$ to be the identity matrix, then the MCG with multiplier matrix $\mathbf{B} = \mathbf{M}_f$ requires only k multiplications to produce a single k-dimensional vector. However, as a vector sequence, the first $k - 1$ rows (or copies of the same MRG) in Eq. 10.9 are simple one-step shifts of each other, which is clearly unsuitable for parallel simulation. The last row, with all the k multiplications, is no more efficient than implementing the MRG directly.

Novel Efficient MCG Implementation of an MRG

It is interesting to observe that choosing the transpose of $\mathbf{M}_f$ (denoted by $\mathbf{M}'_f$) as matrix multiplier $\mathbf{B}$ also gives the same characteristic polynomial since $\det(x\mathbf{I} - \mathbf{M}_f) = \det(x\mathbf{I} - \mathbf{M}'_f)$. By Eq. 10.3,

$$\mathbf{M}'_f = \begin{pmatrix} 0 & 0 & \ldots & 0 & \alpha_k \\ 1 & 0 & \ldots & 0 & \alpha_{k-1} \\ 0 & 1 & \ldots & 0 & \alpha_{k-2} \\ \vdots & \vdots & \ddots & \vdots & \vdots \\ 0 & 0 & \ldots & 1 & \alpha_1 \end{pmatrix}.$$

Then the MCG recursion equation in Eq. 10.7 can be rewritten as

$$\mathbf{X}_i = \begin{pmatrix} X_{i,1} \\ X_{i,2} \\ \vdots \\ X_{i,k} \end{pmatrix} = \begin{pmatrix} \alpha_k X_{i-1,k} \\ X_{i-1,1} + \alpha_{k-1} X_{i-1,k} \\ \vdots \\ X_{i-1,k-1} + \alpha_1 X_{i-1,k} \end{pmatrix} \quad \mathrm{mod}\ p, \ i \geq 1. \tag{10.10}$$

Clearly, the recursion equation in Eq. 10.10 is very efficient to implement because each row requires only one multiplication and at most one addition. Since each α_i is assumed to be nonzero, there is no simple shift relationship between successive variates. However, there are several potential shortcomings for the choice of $\mathbf{B} = \mathbf{M}'_f$: (i) The first row of the generated variate has a simple linear relation as $X_{i,1} = \alpha_k X_{i-1,k}$, not a good generator. Therefore, we recommend discarding the first row from its output. (ii) The remaining $(k-1)$ rows are linear combinations of two previous variates, $X_{i,j} = X_{i-1,j-1} + \alpha_{k-j} X_{i-1,k}$, $j = 2, \ldots, k$, which are similar to the FMRG. With coefficients $\{1, \alpha_{k-j}\}$, they are not great generators either. From the statistical justification point of view as discussed in Deng and Xu (2003), it would be ideal to have "large" and slightly complex coefficients in these generating equations. (iii) The "shifts" among different MRG streams are unknown or hard to determine. Next, we will offer some remedies to ease these shortcomings.

Randomly choose nonzero $t_i \in \mathbb{Z}_p$ for $i = 1, 2, \ldots, k$. If we choose the invertible matrix $\mathbf{T}$ to be

$$\mathbf{T} = \begin{pmatrix} t_1 & 0 & 0 & \cdots & 0 \\ 0 & t_2 & 0 & \cdots & 0 \\ 0 & 0 & t_3 & \cdots & 0 \\ \vdots & \vdots & \vdots & \ddots & \vdots \\ 0 & 0 & 0 & \cdots & t_k \end{pmatrix} \tag{10.11}$$

and consider the multiplier matrix

$$\mathbf{B} = \mathbf{T}\mathbf{M}'_f\mathbf{T}^{-1} = \begin{pmatrix} 0 & 0 & \cdots & t_1/t_k\,\alpha_k \\ t_2/t_1 & 0 & \cdots & t_2/t_k\,\alpha_{k-1} \\ 0 & t_3/t_2 & \cdots & t_3/t_k\,\alpha_{k-2} \\ \vdots & \vdots & \ddots & \vdots \\ 0 & 0 & t_k/t_{k-1} & \alpha_1 \end{pmatrix} \equiv \begin{pmatrix} 0 & \cdots & 0 & \alpha_k^* \\ \gamma_1 & \cdots & 0 & \alpha_{k-1}^* \\ 0 & \gamma_2 & 0 & \alpha_{k-2}^* \\ \vdots & \ddots & \vdots & \vdots \\ 0 & \cdots & \gamma_{k-1} & \alpha_1^* \end{pmatrix}, \tag{10.12}$$

where

$$\alpha_1^* = \alpha_1, \quad \alpha_{i+1}^* = (t_{k-i}/t_k)\alpha_{i+1}, \quad \gamma_i = t_{i+1}/t_i, \quad i = 1, 2, \ldots, k-1,$$

then the MCG recursion equation in Eq. 10.7 can be rewritten as

$$\mathbf{X}_i = \begin{pmatrix} X_{i,1} \\ X_{i,2} \\ \vdots \\ X_{i,k} \end{pmatrix} = \begin{pmatrix} \alpha_k^* X_{i-1,k} \\ \gamma_1 X_{i-1,1} + \alpha_{k-1}^* X_{i-1,k} \\ \vdots \\ \gamma_{k-1} X_{i-1,k-1} + \alpha_1^* X_{i-1,k} \end{pmatrix} \mod p, \quad i \geq 1. \tag{10.13}$$

Using a $\mathbf{T}$ as in Eq. 10.11 generalizes the MCG in Eq. 10.10 to a larger class of MCGs as in Eq. 10.13. Comparing Eqs. 10.10 and 10.13, the MCG in Eq. 10.13 has better a statistical justification than the MCG in Eq. 10.10, and yet the diagonal form of $\mathbf{T}$ preserves the simplicity of the recursion, hence the efficiency. Specifically, excluding the first row, each remaining row in Eq. 10.13 requires two multiplications. In contrast, if there were k nonzero multipliers in $\mathbf{M}_f$, a direct implementation of an MRG-k would require k multiplications. Clearly, the implementation in Eq. 10.13 is dramatically more efficient than the direct implementation. To be more precise, this particular approach necessitates a mere $2k - 1$ multiplications to generate a set of k numbers, while a straightforward implementation described in Eq. 10.9 with k multipliers mandates k^2 multiplications. By opting for random values of $t_i, i = 1, 2, \ldots, k$, we introduce a mechanism for introducing *random* "shifts" across various MRG streams, as opposed to utilizing *fixed* yet unknown "shifts" when α_i's are provided.

Generating by Eq. 10.13 a sequence of $k \times 1$ vectors, $\{\mathbf{X}_i, \ i = 1, 2, \ldots\}$, is equivalent to generating k sequences of random numbers in parallel, which can be assigned to k parallel processors, respectively. However, we recommend always skipping the first element of $\mathbf{X}_i$ in Eq. 10.13 because $X_{i,1}$ for $i \geq 1$ is simply a multiple of the previous number just computed.

10.3.4   Proposed Procedure to Implement an MRG with All Nonzero Terms

When choosing invertible matrix $\mathbf{T} = \text{Diag}(t_1, t_2, \ldots, t_k)$ in Eq. 10.11, abundant selections are available by randomly choosing nonzero t_i for $i = 1, 2, \ldots, k$. To further improve the efficiency, we can choose an integer s_i such that $t_i = 2^{s_i} \mod p$. Therefore, we can rewrite

$$\alpha_1^* = \alpha_1, \quad \alpha_{i+1}^* = (t_{k-i}/t_k)\alpha_{i+1} = (2^{s_{k-i} - s_k})\alpha_{i+1} \mod p,$$

$$\gamma_i = t_{i+1}/t_i = 2^{s_{i+1} - s_i} \mod p, \quad i = 1, 2, \ldots, k-1.$$

It is interesting to observe that, for the modulus $p = 2^{31} - 1$ and $s > 0$, $2^s = 2^{(s \mod 31)} \mod p$ and $2^{-s} = 2^{31-s} \mod p$.

The following procedure summarizes how to implement MRG-k efficiently for d processors in parallel.

1. Apply Algorithm AGM-2 to produce a family of primitive polynomials, say, $f_{r_j}(x)$, $j = 1, 2, \ldots, d$, with all nonzero terms.
2. Implement the following on each of the d processors in parallel.

   a. For simplicity, denote $f_{r_j}(x)$ by $f(x)$ as in Eq. 10.2.
   b. Using the companion matrix $\mathbf{M}_f$ corresponding to $f(x)$, construct the MCG whose multiplier matrix is given in Eq. 10.12 and implement its efficient recursion for $i \geq 1$,

$$
\begin{pmatrix} X_{i,1} \\ X_{i,2} \\ \vdots \\ X_{i,k} \end{pmatrix} = \begin{pmatrix} 2^{s_1-s_k}\alpha_k X_{i-1,k} \\ 2^{s_2-s_1} X_{i-1,1} + 2^{s_2-s_k}\alpha_{k-1} X_{i-1,k} \\ \vdots \\ 2^{s_k-s_{k-1}} X_{i-1,k-1} + 2^{s_k-1-s_k}\alpha_1 X_{i-1,k} \end{pmatrix}
$$

$$
= \begin{pmatrix} \alpha_k^* X_{i-1,k} \\ 2^{s_2-s_1} X_{i-1,1} + \alpha_{k-1}^* X_{i-1,k} \\ \vdots \\ 2^{s_k-s_{k-1}} X_{i-1,k-1} + \alpha_1^* X_{i-1,k} \end{pmatrix} \mod p.
$$

   c. Use the "column-wise" output method to generate k numbers at each recursion, skip the first number, and output the remaining $k - 1$ numbers in sequence.

   Note that $\alpha_1^* = 2^{s_k-1-s_k}\alpha_1 \mod p$ and $\alpha_{i+1}^* = 2^{s_{i+1}-s_k}\alpha_{i+1} \mod p$ for $i = 1, \ldots, k - 1$ can be pre-computed. In addition, $2^s X$ can be implemented efficiently in a computer by a simple logical shift operation. Thus, this implementation is highly efficient for it needs only one multiplication, one logical shift, and one addition for each generated variate.
   Note that we can also use "row-wise" output methods for $k - 1$ processors by assigning them in parallel to $k - 1$ processors respectively. In this case, no need to generate other MCGs unless more than $k - 1$ streams were needed.

10.4   Theory and Relations Between MCG and MRG

Unless stated otherwise, we will assume that any matrix (polynomial) mentioned in this chapter is a $k \times k$ matrix (k-th degree polynomial) with coefficients 0, 1, ..., $p - 1$. The operations such as addition and multiplication are the usual modulo p arithmetic.
   The characteristic function of a matrix $\mathbf{B}$ is defined as

$$
f_{\mathbf{B}}(x) = \det(x\mathbf{I} - \mathbf{B}) \mod p = x^k - \alpha_1 x^{k-1} - \alpha_2 x^{k-2} - \cdots - \alpha_k. \tag{10.14}
$$

Let $f(x) = x^k - \alpha_1 x^{k-1} - \cdots - \alpha_k$ be a primitive polynomial and

$$\mathbf{M}_f = \begin{pmatrix} 0 & 1 & 0 \dots 0 \\ 0 & 0 & 1 \dots 0 \\ 0 & 0 & 0 \dots 0 \\ . & . & . \dots . \\ & & . \dots . \\ 0 & 0 & 0 \dots 1 \\ \alpha_k & \alpha_{k-1} & . \quad . \quad \alpha_1 \end{pmatrix} \qquad (10.15)$$

be the companion matrix of $f(x)$. Using the well-known Cayley-Hamilton theorem, Grothe (1987) showed that the MCG

$$\mathbf{X}_i = \mathbf{B}\mathbf{X}_{i-1} \quad \mathrm{mod}\ p, \quad i \geq 1$$

can be written as

$$\mathbf{X}_i = \alpha_1\mathbf{X}_{i-1} + \alpha_2\mathbf{X}_{i-2} + \cdots + \alpha_k\mathbf{X}_{i-k} \quad \mathrm{mod}\ p, \quad i \geq k. \qquad (10.16)$$

Therefore, each component of the vector sequence generated by a MCG satisfies the same linear recurring relation in Eq. 10.16. Each recurrence sequence with the maximum order defined in Eq. 10.16 is called a M-sequence. The following lemma shows that any two component sequences of a maximum order MCG is a simple shift of each other.

Lemma 10.1 *Two M-sequences differ by a translation if and only if they have the same characteristic function.*

Lemma 10.1 is equivalent to Lemma 1 of Niederreiter (1986), and it has been stated in (Zierler 1959, Sect. 4 on page 39).

Two matrices $\mathbf{B}_1, \mathbf{B}_2\ k \times k$ matrix with coefficients $0, 1, .., p - 1$ are *similar*, $(\mathbf{B}_1 \simeq \mathbf{B}_2)$, if there is a non-singular matrix $\mathbf{T}\ k \times k$ matrix with coefficients $0, 1, .., p - 1$ such that $\mathbf{B}_1 = \mathbf{T}\mathbf{B}_2\mathbf{T}^{-1}$.

Lemma 10.2 *Let p be a prime number, $f_\mathbf{B}(x)$ be the characteristic polynomial of a matrix $\mathbf{B}\ k \times k$ matrix with coefficients $0, 1, .., p - 1$ and $\mathbf{M}_f$ be the companion matrix of a k-th degree polynomial $f(x)$.*

1. *$\mathbf{B}$ has the maximum order $G = p^k - 1$ if and only if $f_\mathbf{B}(x)$ is a k-th degree primitive polynomial.*
2. *For any $\mathbf{B}$ with the maximum order, there exists a non-singular matrix $\mathbf{T}\ k \times k$ matrix with coefficients $0, 1, .., p - 1$ such that $\mathbf{B} = \mathbf{T}\mathbf{M}_{f_\mathbf{B}}\mathbf{T}^{-1}$.*

From Part (2), $\mathbf{B} \simeq \mathbf{M}_{f_\mathbf{B}}$. Therefore, $b = \det(\mathbf{B}) = \det(\mathbf{M}_{f_\mathbf{B}}) = (-1)^{k-1}\alpha_k$. From Theorem 3.18 of Lidl and Niederreiter (1994), we can prove a conjecture by Deng and

George (1990): "A necessary condition for $\mathbf{B}$ $k \times k$ matrix with the maximum order p is that $b = \det(\mathbf{B})$ is a primitive root (mod p)."

Let $\phi(x)$ be the Euler "totient" function of x, which is the number of integers between 1 and x that are relatively prime to x. It is well-known that if $x = p_1^{r_1} p_2^{r_2} \cdots p_c^{r_c}$, then $\phi(x) = x \prod_{i=1}^{c}(1 - \frac{1}{p_i})$. There are exactly $\Phi(p^k - 1)/k$ primitive polynomials of degree k (see, e.g., Knuth 1998). We shall show that with a single $\mathbf{B}$ $k \times k$ matrix with the maximum order p one can generate *all* the k-th degree primitive polynomials in $\mathbb{Z}_p$.

Theorem 10.1 *Let p be a prime number, $\mathbf{B}$ $k \times k$ matrix with the maximum order p. Define*

$$f_r(x) = \det(x\mathbf{I} - \mathbf{B}^r) \quad \mathrm{mod}\ p.$$

1. *$f_r(x)$ is a primitive polynomial if and only if r is relatively prime to $p^k - 1$.*
2. *If r, s are relatively prime to $p^k - 1$, then the following conditions are equivalent:*

    a. *$f_r(x) = f_s(x)$.*
    b. *$r = sp^t \mod (p^k - 1)$, for some integer $t \geq 0$.*
    c. *$\mathbf{B}^r \cong \mathbf{B}^s$.*

    item For any k-th degree primitive polynomial $f(x)$, there exists r relatively prime to $p^k - 1$ such that $f(x) = f_r(x) = \det(x\mathbf{I} - \mathbf{B}^r) \mod p$.

Let $\mathbf{A} = \{X_i, i = 0, 1, 2, \ldots\}$ be a sequence defined in Eqs. 10.14 to 10.16 and $\mathbf{A}^{(r)} = \{X_{ri}, i = 0, 1, 2, \ldots\}$, where r is a positive integer. We will call $\mathbf{A}^{(r)}$ a decimation sequence of M-sequence $\mathbf{A}$. Zierler (1959) showed that every sequence of period $p^k - 1$ may be obtained as a translate of some such $\mathbf{A}^{(r)}$. In particular, if both r, s are relatively prime to $p^k - 1$, then the sequences $\mathbf{A}^{(r)}, \mathbf{A}^{(s)}$ are translations of each other if and only if $r = sp^t$ mod $(p^k - 1)$, for some integer $t \geq 0$.

Note that there is a one-to-one correspondence among

1. M-sequences (translation of sequence considered the same) of period $p^k - 1$,
2. primitive polynomials of degree k,
3. $\mathbf{M}_f$ is $k \times k$ matrix with the maximum order p.

According to Lemma 10.1, two M-sequences differ by a translation if and only if they have the same characteristic function $f(x)$. Let $f(x), \mathbf{M}_f$ be the corresponding characteristic function and the companion matrix for the M-sequence $\mathbf{A} = \{X_i, i = 0, 1, 2, \ldots\}$. Recall that an M-sequence can be generated by its companion matrix $\mathbf{B} = \mathbf{M}_f$ by

$$\mathbf{X}_0, \mathbf{B}\mathbf{X}_0, \mathbf{B}^2\mathbf{X}_0, \ldots..$$

The sequence $\mathbf{A}^{(r)} = \{X_{ir}, i = 0, 1, 2, \ldots\}$ can also be similarly generated with $\mathbf{B} = \mathbf{M}_f^r$. Note the characteristic function of the M-sequence $\mathbf{A}^{(r)}$ is $f_r(x) = \det(x\mathbf{I} - \mathbf{M}_f^r)$.

By taking $\mathbf{B} = \mathbf{M}_f$ in Theorem 10.1, we also provide a general way of generating *all* primitive polynomials of degree k from only one primitive polynomial of degree k. Hence, we have provided a constructive method of generating all primitive polynomials of degree k over Z_p. Lidl and Niederreiter (1994) suggested a general method of determining *all* other primitive polynomial from a known primitive polynomial $f(x)$ over $\mathbb{Z}_p$. Their method involves the calculation of minimum polynomial of θ^r, r is relatively prime to $p^k - 1$, where θ is a primitive root of $f(x)$ in $GF(p^k)$. Golomb (1982) proposed an easier recursive procedure than the method by Lidl and Niederreiter (1994), but it still involves the calculation of a primitive root θ. Clearly, our method is much easier to implement than theirs.

The following result proves that any maximum-order matrix can be generated from a single primitive polynomial $f(x)$.

Corollary 10.1 *Let p be a prime number and $f(x)$ any given primitive polynomial and $\mathbf{M}_f$ be the companion matrix of $f(x)$. Any $\mathbf{A}$ $k \times k$ matrix with the maximum order p can be generated from $f(x)$ by*

$$\mathbf{A} = \mathbf{T}\mathbf{M}_f^r\mathbf{T}^{-1},$$

for some non-singular matrix $\mathbf{T}$ $k \times k$ matrix with coefficients $0, 1, .., p - 1$ and r is relatively prime to $p^k - 1$.

Clearly, if we replace $\mathbf{M}_f$ with any $\mathbf{B}$ $k \times k$ matrix with the maximum order p in Corollary 10.1, the result still holds. That is, we can generate *all* matrices with the maximum order from one known $\mathbf{B}$ with $k \times k$ matrix with the maximum order p by considering

$$\mathbf{T}\mathbf{B}^r\mathbf{T}^{-1}, \quad r \text{ relatively prime to } p^k - 1,$$

where $\mathbf{T}$ $k \times k$ matrix with coefficients $0, 1, .., p - 1$ is non-singular.

The number of non-singular matrices $\mathbf{T}$ $k \times k$ matrix with coefficients $0, 1, .., p - 1$ is

$$N(k, p) = \prod_{i=0}^{k-1}(p^k - p^i), \tag{10.17}$$

and the number of r is relatively prime to $p^k - 1$ for which $\mathbf{B}^r$ producing different similar classes is Theorem 10.1

$$\frac{\phi(p^k - 1)}{k}. \tag{10.18}$$

The number of $\mathbf{A}$ $k \times k$ matrix with the maximum order p is (Niederreiter 1990, Lemma 5).

$$M(k, p) = \frac{\phi(p^k - 1)}{k} \prod_{i=1}^{k-1}(p^k - p^i). \tag{10.19}$$

From Eqs. 10.17 to 10.19, we have

$$N(k, p) \cdot \phi(p^k - 1)/k = (p^k - 1) \cdot M(k, p)$$

Clearly, for a fixed $\mathbf{B}$ $k \times k$ matrix with the maximum order p, r is relatively prime to $p^k - 1$, there are some non-singular matrices $\mathbf{T}$, $\mathbf{S}$ $k \times k$ matrix with coefficients $0, 1, .., p - 1$ such that

$$\mathbf{T}\mathbf{B}^r\mathbf{T}^{-1} = \mathbf{S}\mathbf{B}^r\mathbf{S}^{-1} \quad \mod p. \qquad (10.20)$$

We will prove the following characterization: for a given non-singular $\mathbf{T}$ $k \times k$ matrix with coefficients $0, 1, .., p - 1$, there are exactly $p^k - 1$ non-singular $\mathbf{S}$ $k \times k$ matrix with coefficients $0, 1, .., p - 1$ satisfying Eq. 10.20.

Theorem 10.2 *1. For any nonderogatory matrix $\mathbf{B}$ $k \times k$ matrix with coefficients*
$0, 1, .., p - 1$,

$$\mathbf{T}\mathbf{B}\mathbf{T}^{-1} = \mathbf{S}\mathbf{B}\mathbf{S}^{-1} \quad \mod p,$$

if and only if

$$\mathbf{S} = \mathbf{T}\, g(\mathbf{B}),$$

where $g(x)$ is a polynomial with $0 \leq \deg g \leq k - 1$.
2. Furthermore, if $\mathbf{B}$ is $k \times k$ matrix with the maximum order p, then for each non-singular
matrix $\mathbf{T}$ $k \times k$ matrix with coefficients $0, 1, .., p - 1$, there are exactly $p^k - 1$ non-
singular matrices $\mathbf{S}$ satisfying (3.2.9).

There are several potential flaws to all MCG:

1. According to Lemma 10.1, any two component sequences of a MCG with the maximum order is a simple translation of each other.
2. Similar to the observation by Marsaglia (1985) with $p = 2$, there is always a matrix $\mathbf{T}$, according to Lemma 10.2, such that the transformed sequence $\mathbf{Y}_i = \mathbf{T}^{-1}\mathbf{X}_i$ satisfying

$$\mathbf{Y}_{i+1} = \mathbf{M}_f\mathbf{Y}_i \quad \mod p.$$

Therefore, the first $k - 1$ components of $\mathbf{Y}_{i+1}$ is a simple shift from previous $\mathbf{Y}_i$. Clearly, this is not a desirable property for a RNG.

Another interpretation of the above observation is that MCG makes k-duplicates of one MRG; thus MCG does not provide more "randomness" property than the MRG itself.

Remark

The general maximum-period MRGs with many nonzero terms have all the qualities of a good random number generator except that its direct implementation is inefficient and parallel implementation becomes increasingly difficult as the order k increases. Current maximum-period MRGs utilize certain simple structures in the multipliers to gain a significant increase in their efficiency. However, there is a tradeoff—these MRGs with few nonzero terms have a larger spectral distance than that of an MRG with many nonzero terms.

The proposed automatic generation algorithm AGM-2 can find numerous general MRGs from one with a simple structure. There is no practical limit on the number of MRGs that can be produced by Algorithm AGM-2 and each MRG has an extremely long period length, equidistribution property up to order k, and, with many nonzero terms, it is expected to perform well on spectral test. This abundance source of general MRGs had remained untapped without an efficient implementation until now.

Part IV
Secure Random Number Generators

Classical Secure Generators

11

Stream ciphers produce a sequence of ciphertext (in bits, bytes, words or larger sizes) sequentially by adding a variate from a key stream (of the same size) to a sequence of plaintext. The key stream sequence generated by a synchronous stream cipher depends only on the key, not on the plaintext or ciphertext. Therefore, it is clear that the security of a stream cipher completely depends on how "good" the generated key stream sequence is. If one could generate a sequence of truly random variates for the key stream, then the resulting sequence of ciphertext would behave like a random sequence of variates. Consequently, it would be hard for attackers to decrypt from the ciphertext without knowing the values of the generated key stream. It is clear that no generators based on deterministic algorithms can claim to generate a truly random sequence of uniformly distributed random variates. Therefore, it is essential to design "good" pseudo-random number generators (PRNGs) to generate the key stream for a stream cipher. There are several desirable properties for a "good" random number generator: (i) long period length so that the key stream sequence would not repeat quickly, (ii) forward and backward unpredictability so that attackers could not predict the future variates from observing a number of past generated variates, (iii) great support of statistical/distributional properties for excellent empirical performances, (iv) great efficiency—fast generating speed and less memory space/gate count (mainly for hardware), and (v) simplicity, extendibility, and flexibility of the design.

Due to different objectives and concerns, two very different types of pseudo-random number generators (PRNGs) have been developed in the literature and used in practice for applications in computer simulation and computer security. For many computer simulation applications, it is common to use PRNGs to produce a sequence of variates that is very hard to distinguish from a sequence of (truly) random numbers. However, most of the proposed generators are linear generators and the linearity makes the generated sequences easily

L. Deng et al., *Random Number Generators for Computer Simulation and Cyber Security*, Synthesis Lectures on Mathematics & Statistics, https://doi.org/10.1007/978-3-031-76722-7_11

predictable from just a few past values. In contrast, the PRNGs developed for computer security applications emphasize the issue of security.

For security applications, many pseudo-random bit generators (PRBGs), also known as deterministic random bit generators (DRBGs), have been developed. Barker and Kelsey (2012) published NIST's recommendation and specification on approved DRBG algorithms, including those based on hash functions, block cipher algorithms (AES and Triple DES), and a number theory problem that is expressed in elliptic curve technology. The popular RC4 stream cipher designed by Rivest is a byte-oriented stream cipher because it outputs only 8 bits (one byte) at each iteration. Another popular generator is the shrinking generator proposed by Coppersmith et al. (1994), which was shown to be vulnerable to multiple attacks including the correlation attack (Golic 2001). Nonetheless, the shrinking generator is an interesting example of using one LFSR (Linear Feedback Shift Register) acting as a (random) selector to choose or skip the output bit produced by another LFSR. Many other generators consider the strategy of applying some kind of non-linear filtering processes on multiple LFSRs (see, e.g., Menezes et al. 1996; Robshaw and Billet 2008).

An MRG (Multiple Recursive Generator) of order k computes its next variate by a linear combination of the past k variates, over a finite field of a large prime modulus m. MRG includes LFSR as a special case with modulus 2. MRGs have strong statistical justifications and excellent empirical performances. Another type of PRNG having great theoretical properties is MT19937, proposed by Matsumoto and Nishimura (1998), which is currently the most popular 32-bit PRNG and it has a period length of $2^{19937} - 1 \approx 10^{6001}$ with the dimensions of equi-distribution up to 623.

Neither MRGs nor LFSRs (including MT19937) are secure generators because one can compute their next value from a sequence of observed past values by solving a system of linear recurrence equations. See, for example, Stinson (2006) for cryptanalysis of the LFSR and Lidl and Niederreiter (1994) for that of the MRG. An MRG of order k requires up to k past known variates (in correct order) to determine its next variate. Therefore, to make it harder for attackers to predict the next variate, Deng et al. (2018) proposed to scramble or hide (or both) the generated values by shuffling or mixing with other varieties.

11.1 Strength and Weakness of the Linear Generators

Strength

Multiple recursive generators of large order k have extremely long period length and high dimensional equi-distribution property. While general MRGs can be inefficient, DX-k generators are efficient as LCGs and DX generators have been shown to pass the stringent TestU01 in L'Ecuyer and Simard (2007). In addition, DX generatorsand some MRGs can be implemented in parallel as proposed in Deng et al. (2009a). Currently, the most popular generator is MT19937 proposed by Matsumoto and Nishimura (1998) which is a variant of GFSR and it is currently the default generator used in the R package with a period length of $2^{19937} - 1 \approx 10^{6001}$ with the dimensions of equi-distribution up to 623. Therefore, these liner generators are suitable for computer simulation applications. For more details, please see a review in Deng and Bowman (2017) on the recent development.

Weakness

The main reason that linear generators are not recommended is because of the key weakness in their predictability. Due to its linear structure, it is somewhat simple to predict linear generators. For example, we can break the system (solve parameters or predict its future values) with a small portion of the sequence of variates observed. The key idea is that one can solve a system of linear equations. In fact, we can easily solve the linear equations when the modulus m is known. Even if modulus m is unknown, we can break the linear system via the following examples and theorems.

Example 11.1 (*Predicting LCGs: m known*)

If modulus m is known, then it is easy to find B with only a pair of variates needed:

$$X_i = B X_{i-1} \mod m, \quad B = (X_{i-1})^{-1} X_i \mod m.$$

Similar (but harder) problems for MRG and LFSR are found by solving a system of linear equations. We will start the straightforward procedure to break an LCG when m is unknown as below:

Theorem 11.1 (Predicting LCGs: m unknown) *For any $i \neq j$, let*

$$D_{ij} = \left\| X_{i-1} X_j - X_i X_{j-1} \right\|$$

Note that $D_{ij} \mod m = 0$, for any pair of i and j because $X_i = B X_{i-1} \mod m$ and $X_j = B X_{j-1} \mod m$. Therefore, $m = \gcd(D_{i,i+1}, i \in S)$, for a set S.

Example 11.2 (*Predicting LCGs: m unknown*)

Let's use a simple example with observed $LCG(B, m)$ sequence: 17, 77, 35, 39, 87, 28, 82, 95, 124 with *unknown m and B*.

1. Compute values of $D_{i,i+1} = \left\| X_{i-1} X_{i+1} - X_i^2 \right\|$:

   - $5334 = |17 \times 35 - 77^2|$
   - $1778 = |77 \times 39 - 35^2|$
   - $1524 = |35 \times 87 - 39^2|$
   - ...

2. Compute the greatest common divisor of $D_{i,i+1}$: $\gcd(5334, 1778, 1524, \ldots) = 127$, that is, $m = 127$
3. Solve for B with $m = 127$: $B = (17)^{-1}77 \bmod 127 = 15 * 77 \bmod 127 = 12$.

We can use a similar (but harder) strategy to break MRGs and LFSRs. There are more advanced procedures to break the linear generators in the literature (see, e.g., Boyar 1989).

Neither MRGs nor LFSRs (including MT19937) are secure generators because one can compute their next value from a sequence of observed past values by solving a system of linear recurrence equations (see, e.g., Stinson 2006; Lidl and Niederreiter 1994, for crypt analysis of the LFS Rand for that of the MRG, respectively). An MRG of order k requires up to k past known variates (in correct order) to determine its next variate. Therefore, to make it harder for attackers to predict the next variate, Deng et al. (2018) proposed to scramble or hide (or both) the generated values by shuffling or mixing with other variates.

For applications in computer security, one possible solution is to modify the linear generators with a series of non-linear transformations to break the linearity structure while inheriting some of its nice properties of long period length, efficiency, distributional property, and great empirical performance. At the same time, we hope to enhance the security property so that they can be also used for computer security applications.

In the next subsection, we will give a general discussion on stream cipher and its connection with (secure) random number generators.

11.2 Stream Cipher and Secure RNG

Stream ciphers produce a sequence of ciphertext (in bits, bytes, words, or larger sizes) sequentially by adding a variate from a key stream (of the same size) to a sequence of plaintext. The key stream sequence generated by a synchronous stream cipher depends only on the key, not on the plaintext or ciphertext. Therefore, it is clear that the security of a stream cipher completely depends on how "good" the generated key stream sequence is. If one could generate a sequence of truly random variates for the key stream, then the resulting sequence of ciphertext would behave like a random sequence of variates. Consequently, it would be hard for attackers to decrypt the ciphertext without knowing the values of the generated key stream. It is clear that no generators based on deterministic algorithms can claim to generate a truly random sequence of uniformly distributed random variates. Therefore, it is essential to design "good" pseudo-random number generators (PRNGs) to generate the key stream for a stream cipher. There are several desirable properties for a "good" random number generator:

(i) long period length so that the key stream sequence would not repeat quickly,
(ii) forward and backward unpredictability so that attackers could not predict the future variates from observing a number of past generated variates,
(iii) great support of statistical/distributional properties for excellent empirical performances, (iv) great efficiency—fast generating speed and less memory space/gate count (mainly for hardware), and
(iv) simplicity, extendibility, and flexibility of the design.

Secret Keys and IVs

For stream ciphers with the same input keys, it is important to produce a different stream sequence with different initial values (IVs) which may not be kept secret.

Vernam's One-Time Pad Encryption

The theory behind Vernam's one-time pad encryption: "perfect secrecy" Shannon (1948, 1949).

Theorem 11.2 *Let $P_1, P_2, \ldots$ be any sequence of plain text with possible values from $\mathbb{Z}_m = \{0, 1, \ldots, m - 1\}$. Suppose that we can generate a sequence of random variates $X_1, X_2, \ldots$ that are i.i.d. from a uniform distribution over $\mathbb{Z}_m$. Set (secret) cipher text $C_i = P_i + X_i$ mod m. Then, $C_1, C_2, \ldots$ is a random i.i.d. sequence from uniform distribution over $\mathbb{Z}_m$.*

A few notes about Vernam's one-time pad encryption in practice:

1. Given only the cipher text $C_i = P_i + K_i$ mod m, one cannot recover/"guess" the plain text P_i.
2. Given both cipher text (C_i) and plain text (P_i), K_i can be easily found. Hence, the same K_i sequence can *only be used once*.
3. It is common to use a pseudo-random algorithm to produce such K_i sequence. Such an algorithm should have unpredictability property in the sense that no matter how many variables in the sequence have been observed in the past, one cannot predict its future value.

It is interesting to note that there is a connection between One-time pad and statistics which is the "combination" of two PRNGs:

$$U_i = X_i + Y_i \mod m, \quad Y_i = U_i - X_i \mod m$$

In particular, the combination procedure can mask the values of X_i (K_i) and Y_i (P_i) and it can improve the distributional property over the components PRNGs. When $m = 2$ or $m = 2^{32}$, we can replace the addition with XOR in the above equation as

$$C_i = K_i \oplus P_i, \quad P_i = C_i \oplus K_i. \tag{11.1}$$

11.3   Stream Cipher Versus Block Cipher

A block cipher encrypts a given plaintext divided into fixed-size blocks (either 64 or 128 bits) using the same key and a cryptographic algorithm. In contrast, a stream cipher using some algorithm with input key vector (K) to produce a sequence of K_i, $i = 1, 2, \ldots$ to encrypt P_i as in Eq. 11.1 to produce a sequence of cipher text C_i, where m could be 2, 2^8 or even 2^{32}. Compared to a block cipher, the encryption/decryption procedure as in Eq. 11.1 for a stream cipher is much simpler. In addition, a stream cipher can take various IVs and the same key (K) to produce a different sequence of K_i, $i = 1, 2, \ldots$.

11.3.1 Common Attack Models

The attack model specifies the information available to the adversary when he mounts his attack. The most common types of attack models are enumerated as follows.

1. **Ciphertext only attack**: The opponent possesses a string of ciphertext, y.
2. **Known plaintext attack**: The opponent possesses a string of plaintext, x, and the corresponding ciphertext, y.
3. **Chosen plaintext attack**: The opponent has obtained temporary access to the encryption machinery. Hence he can choose a plaintext string, x, and construct the corresponding ciphertext string, y.
4. **Chosen ciphertext attack**: The opponent has obtained temporary access to the decryption machinery. Hence he can choose a ci phertext string, y, and construct the corresponding plaintext string, x.

In each case, the objective of the adversary is to determine the key that was used. This would allow the opponent to decrypt a specific "target" ciphertext string, and further, to decrypt any additional ciphertext strings that are encrypted using the same key.

11.3.2 Classical Cryptanalysis and Kerckhoffs's Principle

Several common cryptanalysis techniques for stream ciphers are proposed, and we briefly summarize them as described in Bokhari et al. (2012).

1. **Exhaustive/brute force key search attack**: Not effective when the key sizes are large or for "provably secure ciphers" (e.g. one-time pad cipher)
   In an exhaustive key search or brute force attack, the cryptanalyst tries all possible keys to decrypt a ciphertext. Except for provably secure ciphers [1], this method can be used against any cryptographic algorithm, including stream ciphers, Note that a provably secure cipher is not practically feasible, e.g., a one-time pad.
2. **Side channel analysis attack**: Based on the physically observable characteristics during execution such as the power and microprocessor time required for execution, electromagnetic radiation, heat dissipation, and noise of the system, etc.
3. **Time memory trade-off attacks**: Improvement to the exhaustive key search attack that trades off computational time against memory complexity with lower complexity than lookup table and/or an online complexity lower than exhaustive key search.
4. **Distinguishing attacks**: Good stream ciphers should require the keystream generation to produce a random and uniform sequence. A distinguishing attack tries to identify the possible relations between internal state variables and output keystream.
5. **Algebraic attack**: Model a cryptographic system in terms of algebraic equations by finding the set of algebraic equations that relate the initial state with the output keystream, then observed keystream bits are substituted into the equations to collect possible keystream bits information and determine the initial state and the secret key.
6. **Correlation attacks**: Extract some information about the initial state from the output keystream by exploiting the weaknesses in the simple combination function of the design, especially for cipher design based on feedback shift registers.
7. **Guess and determine attacks**: Guess a small part of the internal state and try to recover the full value of internal states with the observed keystream and check the correctness of the guessed values from the keystream generated using the guessed values.
8. **Linear masking attacks**: It is a form of the previously stated Guess and Determine attack. A non-linear characteristic is identified with a linear process and some missing linear combinations which are used to find the traces of distinguishing property.
9. **Related key attack**: Useful for ciphers with a constant rekeying strategy with either new key or IV (initialization vector), especially for those designs without sufficient non-linearity from inputs to the internal states.
10. **Divide and conquer attack**: Divide a cipher system into several components and try to attack the most vulnerable components first.

Finally, we also mention an important principle promoted by Kerckhoffs (1883): A cipher should be *secure* when the enemy cryptanalyst *knows all details* of the enciphering process and deciphering process *except* for the value of the *secret key*. This is a sharp contrast with another principle: "*security through obscurity*". Kerckhoff's Principle is widely embraced by cryptographers.

11.4   Classical Secure Generators

Due to different objectives and concerns, two very different types of pseudo-random number generators (PRNGs) have been developed in the literature and used in practice for applications in computer simulation and computer security, respectively. For many computer simulation applications, it is common to use PRNGs to produce a sequence of variates that is very hard to distinguish from a sequence of (truly) random numbers. However, most of the proposed generators are linear generators and the linearity makes the generated sequences easily predictable from just a few past values. In contrast, the PRNGs developed for computer security applications put the major emphasis on the security issue.

For security applications, many pseudo-random bit generators (PRBGs), also known as deterministic random bit generators (DRBGs), have been developed. Barker and Kelsey (2012) published NIST's recommendation and specification on approved DRBG algorithms, including those based on hash functions, block cipher algorithms (AES and Triple DES), and a number theory problem that is expressed in elliptic curve technology. The popular RC4 stream cipher designed by Rivest is a byte-oriented stream cipher because it outputs only 8 bits (one byte) at each iteration. Another popular generator is the shrinking generator proposed by Coppersmith et al. (1994), which was shown to be vulnerable to multiple attacks including the correlation attack Golic (2001). Nonetheless, the shrinking generator is an interesting example of using one LFSR (Linear Feedback Shift Register) acting as a (random) selector to choose or skip the output bit produced by another LFSR. Many other generators consider the strategy of applying some kind of non-linear filtering processes on multiple LFSRs, (see, e.g., Menezes et al. 1996; Robshaw and Billet 2008).

11.5   Blum Blum Shub (BBS) Genertor

For security applications, we usually choose a function T so that it can hide as many bits of information as possible. Stinson (2006) discussed several random bit generators (PRBGs) with $T(x) = x \mod 2$, which extract only the least significant bit from x regardless of the size of x. While this kind of transformation can hide most of the bit information in x (thus more secure), it significantly slows down the generating speed.

Blum-Blum-Shub (BBS) PRBG with $f(x) = x^2$ and $T(x) = x \mod 2$. The modulus $m = p \times q$, p and q are two large private prime numbers. Both p and q are congruent to 3 mod 4. This generator is first proposed by Blum et al. (1986). To speed up the bit generation while maintaining its security, we can extract r least significant bits from x with $T(x) = x$

mod 2^r, for $r < \log_2(\log_2(m))$. For $m < 2^{1024}$, we can extract $r < 10$ bits at a time, (see Stinson 2006, page 334).

The two large primes, p and q, should both be congruent to 3 (mod 4) because this can guarantee (mathematically) that each quadratic residue has one square root which is also a quadratic residue. For a rigorous security analysis, there is a mathematical proof reducing its security to the computational difficulty of factoring $m = p \times q$. When the primes are properly chosen, and some lower-order bits of each sequence are output, the output bits from random should be at least as difficult as solving the quadratic residue problem modulo m.

The performance of the BBS random-number generator depends on the size of the modulus m and the number of bits per iteration.

Specifically, BBS generators proposed by Blum et al. (1986).

$$X_i = X_{i-1}^2 \mod m, \quad m = p \times q.$$

Deliver *only one bit* (*very slow*):

$$Y_i = X_i \mod 2.$$

where

- p and q are *secret large primes* (of the order 2^{1024}) with $p \mod 4 = 3$ and $q \mod 4 = 3$.
- We need p and q *large enough (of the order 2^{1024})* so that it is possible to factor $m = p \times q$.
- System can be broken with *only one known X_i*.

There are two more PRBG based on the difficulty of mathematical property proposed in the literature:

1. Discrete Logarithm PRBG with $f(x) = \alpha^x$ and $T(x) = \lfloor \frac{2x}{m} \rfloor$: Here m is a large prime number and α is a primitive element modulo m. This generator is based on the same security assumption as that in the Discrete Logarithm p
2. RSA PRBG with $f(x) = x^b$ and $T(x) = x \mod 2$: This generator is based on the same security assumption as that in the RSA cryptosystem proposed by Rivest et al. (1978), in which the modulus $m = p \times q$, p and q are two secret prime numbers, and b is the public key in RSA setting.

The security of the above PRBGs requires the modulus m to be as large as a number with 1024 binary digits, which is equivalent to about 309 digits in decimal. Its generating speed is slow because the modulus m involved is large and just one bit is generated during each iteration. In most simulation applications, we need to generate a pseudo-random number that consists of many binary bits. While it is possible to concatenate these generated bits into a number, it is clearly very time-consuming. Therefore, those PRBGs are not appropriate for simulation applications due to the slow generating speed and no assurance of the uniform distribution property. Even for some cryptographic applications, all of the PRBGs discussed above can be too slow.

11.6   Clock-Controlled Generators

Due to the linearity structure, simple LFSRs are not appropriate for security applications. Therefore, it is common to transform the states of an LFSR or combine several LFSRs into a non-linear Boolean generator. One can classify these proposals into two types: regularly clocked and irregularly clocked. For the regularly clocked generators, the movement of data in all component LFSRs is controlled by the same clock. The first type of generators includes non-linear combination generators (combining several LFSRs via a non-linear function) and non-linear filter generators (combining bits from a single LFSR via a non-linear function). The second type of generators uses the output of one LFSR as a "clock" to control the output of other LFSR(s). There are three popular clock-controlled generators: (1) the Geffe generator proposed by Geffe (1973), (2) the alternate step generator proposed by Günther (1988), and (3) the shrinking generator proposed by Coppersmith et al. (1994) (See Menezes et al. 1996, Chap. 6 for a detailed description). Since we will compare our proposed generator with some clock-controlled generators, we briefly describe them below.

The Geffe generator is based on three LFSRs, two are the baseline generators and the third one is the selector to control the (random) selection of bits generated from one of the two baseline generators. The alternate step generator is similar to the Geffe generator except no bits are discarded. The shrinking generator is based on two LFSRs, one serves as the baseline generator and the other as the selector to control the (random) selection of bits generated from the baseline generator.

In summary, ASG is based on three linear-feedback shift registers. Its output is a combination of two LFSRs which are stepped (clocked) in an alternating fashion, depending on the output of a third LFSR. LFSRs have good distribution and they are simple to implement. However, they cannot be directly because their output can be predicted easily. An ASG comprises three linear-feedback shift registers and the output of one of the registers decides which of the other two is to be used. The output is the exclusive OR of the last bit produced by LFSRs.

11.7   Shrinking Generators

The shrinking generator is to be used in a stream cipher proposed by Coppersmith et al. (1994). The shrinking generator uses two LFSRs. One LFSR is used to generate output bits, while the other LFSR controls their output. If the control bit is 1, then the baseline bit is output; if the control bit is 0, the baseline bit is discarded, and nothing is output. This has the disadvantage that the generator's output rate varies irregularly, and in a way that hints at the state of control LFSR; this problem can be overcome by buffering the output. The random sequence generated by LFSR can not guarantee the unpredictability in a secure system and various methods have been proposed to improve its randomness.

Example 11.3 (*Shrinking generator*)

Given two sequences, $X_1, X_2, \ldots$ and $Y_1, Y_2, \ldots$, produced by two LFSRs, we can pick the X_i's for which the corresponding $Y_i = 1$. In other words, we form a subsequence of X_i's by skipping those X_i with $Y_i = 0$. The shrinking generator is fast but part (about half, on average) of the original sequence is known. Therefore, it is vulnerable to the correlation attack proposed by Golic (2001).

11.8   Rivest Cipher 4 (RC4)

RC4 is one of the most studied stream ciphers which was designed by R. Rivest, one of three inventors of RSA. RC4 is used in many applications: (a) Netscape's Secure Sockets Layer (SSL) protocol, (b) WEP, IEEE 802.11 wireless networking security standard. However, no "rigorous justification" (e.g. BBS) has been provided and some weaknesses/attacks have been found.

RC4 is a well-known byte-oriented software-designed stream cipher. During the initialization stage, the S-table is a shuffled table (based on the user-input key values) of $S[i] = i$, $i = 0, \ldots, 255 (= 2^8 - 1)$. The generation stage of RC4 can be shown in the following diagram (Fig. 11.1).

For RC4, it is common to set $N = 2^8$. The security strength of RC4 relies on the unpredictability of the hidden "internal states" of I, J, and the S-table. The values for these "internal states" are "moving targets" becauseof the continual updates on I, J,

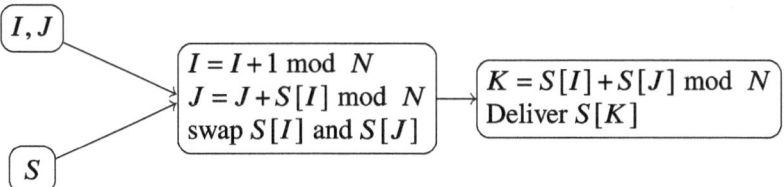

Fig. 11.1 Diagram of RC4

and "random" permutation of the S-table (via the swapping of $S[I]$ and $S[J]$) at each iteration. Several problems with RC4 have been identified, including correlations between consecutive bytes (Fluhrer and McGrew 2000), classes of weak keys (Roos 1995; Wagner 1995), and bias in the second byte of RC4 (Mantin and Shamir 2001). Mironov (2002) provides descriptions of attacks on RC4.

eSTREAM Ciphers

12

The eSTREAM was a multi-year effort launched in 2004 by the ECRYPT (European Network of Excellence in Cryptology) with the goal to "promote the design of efficient and compact stream ciphers suitable for widespread adoption". according to a recent eSTREAM report entitled "eSTREAM Portfolio in 2012" (http://www.ecrypt.eu.org/stream/), There are two "profiles" of the eSTREAM finalist: Profile 1. Stream ciphers for software applications with high throughput requirements.

Rabbit: proposed by Boesgaard et al. (2003),
HC-256/HC-128:   proposed by Wu (2004),
Salsa20/ChaCha: proposed by Bernstein (2008b),
SOSEMANUK: proposed by Berbain et al. (2008),

and Profile 2. Stream ciphers for hardware applications with restricted resources such as limited storage, gate count, or power consumption.

Grain v1:   proposed by Hell et al. (2007),
MICKEY 2.0:   proposed by Babbage and Dodd (2008),
Trivium: proposed by De Canniere and Preneel (2008).

For more information, see Robshaw and Billet (2008). The main evaluation criteria are as follows:

1. Security.
2. Performance when compared to AES in some appropriate mode (e.g., counter mode) and other competing submissions.

© The Author(s), under exclusive license to Springer Nature Switzerland AG 2025   171
L. Deng et al., *Random Number Generators for Computer Simulation and Cyber Security*, Synthesis Lectures on Mathematics & Statistics,
https://doi.org/10.1007/978-3-031-76722-7_12

3. Justification and supporting analysis.
4. Simplicity and flexibility.
5. Completeness and clarity of the submission.

12.1   Profile 1: Software Ciphers

We briefly describe the key features of the finalist, especially their software ciphers. Later, we will discuss some of its "enhancements" to these software ciphers.

12.1.1 HC-256 and HC-128

One of four finalists in the eSTREAM Portfolio is HC-256, proposed by Wu (2004). HC-128 is a smaller variant of HC-256; see http://www.ecrypt.eu.org/stream. HC-128 has two secret tables, P and Q, each with 512 32-bit integers (words). At each iteration, HC-128 (a) updates the P/Q table with a non-linear function and (b) outputs one 32-bit word by mixing some nonlinear functions of entries. According to (Paul and Maitra 2011, Chap. 10, page 233), "HC-128 can be considered as the next level of evolution in the design of RC4-like software stream ciphers."

We briefly describe some popular stream ciphers such as HC-128/HC-256 and its related ciphers. One useful source of information can be found in eSTREAM, https://cr.yp.to/streamciphers.html.

For HC-128, two large tables are required to obtain a long period length and a randomness property. Even though the exact period length of HC-128 is unknown, but it is believed to be of the order of 2^{256}. The tables are usually created (hopefully randomly) based on the secret user key and initialization vector (**IV**) through a series of scrambling procedures (the so-called key scheduling algorithm). Once the table(s) are created, at each iteration, a keystream variate is generated internally via a complicated nonlinear transformation within the entries in the table(s) and continuous updates of the table(s). This is due to the fact that these ciphers do not have external generator(s) to continuously feed and update the table(s) after the initial table(s) have been sufficiently mixed and randomized.

Recall: Notation

Unless explicitly specified, we assume all the arguments and the returned values of any function are integers. For two integers x and y, we define:

- $x \& y$ to be the number corresponding to the bit-wise logical and of their binary representations,

- $x \oplus y$ to be the bit-wise exclusive or,
- $x \boxplus y$ to be $(x + y) \mod 2^{32}$,
- $x \boxminus y$ denote $x - y \mod 512$,
- $x \| y$ the concatenation of the bit strings of x, y,
- $x \ggg n$ denote the right rotation operator, $((x \gg n) \oplus (x \ll (32 - n)))$,
- $x \lll n$ the left rotation operator $((x \ll n) \oplus (x \gg (32 - n)))$.
- $x \gg d$ represents the logical right shift operation,
- $x \ll d$ be the logical left shift operation of d bits in the binary representation of x,
- For an integer m, we denote $\mathbb{Z}_m = \{0, 1, \ldots, m - 1\}$.

HC-128 has two secret tables, P and Q, each with 512 32-bit integers. At each iteration, HC-128 updates an element of one table with a non-linear feedback function of several entries in the same table and generates one 32-bit output by mixing some nonlinear functions of entries from both tables. While it is fairly efficient during the generation stage, its initialization stage requires a long clock cycle. According to Wu (2004), it is difficult to predict the exact period of the cipher, but it is estimated to be much greater than 2^{256}. In order to understand the procedure of the initialization, and its modification for eHC, we reproduce some of the details here.

Operations/Functions Used

Let **K** denote the 128-bit key of HC-128, and **IV** a 128-bit initialization vector. The generated keystream is denoted s and s_i the 32-bit output of the i-th step. Thus, $s = s_0 \| s_1 \| s_2 \| \cdots$. For a 32 bit word x, let $x = x_3 \| x_2 \| x_1 \| x_0$, where x_0, x_1, x_2 and x_3 are each 4 bytes. The following mixing functions are used.

1. $f_1(x) = (x \ggg 7) \oplus (x \ggg 18) \oplus (x \gg 3)$.
2. $f_2(x) = (x \ggg 17) \oplus (x \ggg 19) \oplus (x \gg 10)$.
3. $g_1(x; y; z) = ((x \ggg 10) \oplus (z \ggg 23)) \boxplus (y \ggg 8)$.
4. $g_2(x; y; z) = ((x \lll 10) \oplus (z \lll 23)) \boxplus (y \lll 8)$.
5. $h_1(x) = Q[x_0] \boxplus Q[256 + x_2]$.
6. $h_2(x) = P[x_0] \boxplus P[256 + x_2]$.

It is useful to explain the features of the above mixing functions used:

1. $f_1(x)$ and $f_2(x)$ are mixing/coupling two bit-wise rotation (with $\ggg$ operation) and one shift operation (with $\gg$ operations) of a 32-bit integer x. Note that $f_1(x)$ and $f_2(x)$ are similar to the one used in SHA-256. Since some of the generated variables are 32-bit positive integers with 0 as its sign bit. The fixed sign bit can be seen as a limitation or a problem, but bit-wise rotation and mixing helps overcome this problem.

2. $g_1(x; y; z)$ and $g_2(x; y; z)$ are adding on three 32-bit integer x, y and z with bit-wise rotation (with $\ggg$ or $\lll$ operation).
3. $h_1(x)$ ($h_2(x)$) is adding two entries, indexed randomly by the first byte, the 8-bit x_0, of 32 bit integer $x = x_3\|x_2\|x_1\|x_0$ from the first half of $Q(P)$ table and another from the second half $Q(P)$ table (indexed by $256 + x_2$).

The above functions are typical operations, known as ARX (Addition, Rotation and XOR) operation, which are useful to make non-linear transformations (mostly with rotation) and mixing/addition can hide the individual values and also improve the distributional property as combination generator. See, for example, Deng and George (1990). Random table updates/combination (e.g. $h_1(x)$ and $h_2(x)$) can be useful to resist the powerful algebraic attacks.

HC-128 Initialization

The initialization algorithm is the following.

Algorithm 12.1: HC-128 initialization algorithm

Let $K = K_0\|K_1\|K_2\|K_3$ and $IV = IV_0\|IV_1\|IV_2\|IV_3$;
Let $K_{i+4} = K_i$, and $IV_{i+4} = IV_i$ for $0 \leq i < 4$. K and IV are expanded into an array $W_i (0 \leq i \leq 1279)$ as:

$$W_i = \begin{cases} K_i, & 0 \leq i \leq 7 \\ IV_{i-8}, & 8 \leq i \leq 15 \\ f_2(W_{i-2}) \boxplus W_{i-7} \boxplus f_1(W_{i-15}) \boxplus W_{i-16} \boxplus i, & 16 \leq i \leq 1279 \end{cases}$$

Update the tables P and Q with the array W:

$$P[i] = W_{i+256}, Q[i] = W_{i+768}, \quad \text{for } 0 \leq i \leq 511$$

Run the cipher 1024 steps using outputs to replace the table elements as:
for $i = 0$ *to 511* **do**
    $P[i] = (P[i] + g_1(P[i \boxminus 3]; P[i \boxminus 10]; P[i \boxminus 511])) \oplus h_1(P[i \boxminus 12])$.
    $Q[i] = (Q[i] + g_2(Q[i \boxminus 3]; Q[i \boxminus 10]; Q[i \boxminus 511])) \oplus h_2(Q[i \boxminus 12])$.
end

Here are a few comments about the above initialization procedure:

1. The HC-128 initialization steps 1—2 expanded the four 32-bit K and four 32-bit IV into a array of $1280(256 + 512 + 512)$ 32-bit $W[i]$ sequentially using the some ARX operations.

2. Step 3 simply skipped 256 32-bit of array $W[\cdot]$ and copied the remaining 512 32-bit words into P and Q, respectively.

3. To ensure enough randomness and uniformity of P/Q, steps 4—8 performed additional ARX operations to mix (sequentially) the 1024 entries of P/Q tables.

4. Clearly, the initialization procedure is long because it would require a long ARX operation to yield P/Q tables with desirable uniformity and randomness properties which was needed for its keystream generation procedure. However, no strong theoretical justification such goals can be achieved for any kind of Key/IV selected.

HC-128 Keystream Generation

The keystream generation is done by the following algorithm.

Algorithm 12.2: HC-128 keystream generation algorithm

$i = 0$
repeat
    $j = i$ mod 512;
    if *(i mod 1024) < 512* **then**
        $P[j] = P[j] + g_1(P[j \boxminus 3]; P[j \boxminus 10]; P[j \boxminus 511])$
        $s_i = h_1(P[j \boxminus 12]) \oplus P[j]$
    end
    else
        $Q[j] = Q[j] + g_2(Q[j \boxminus 3]; Q[j \boxminus 10]; Q[j \boxminus 511])$
        $s_i = h_2(Q[j \boxminus 12]) \oplus Q[j]$
        $i = i + 1$
    end
until *until enough keystream bits are generated*;

A few comments about the above keystream generation procedure:

1. The HC-128 initialization steps 1–2 expanded the four 32-bit K and four 32-bit IV into a array of $1280(256 + 512 + 512)$ 32-bit $W[i]$ sequentially using some ARX operations.

2. Step 3 simply skipped 256 32-bit of array $W[\cdot]$ and copied the remaining 512 32-bit words into P and Q, respectively.

3. To ensure enough "randomness" and "uniformity" of P/Q, steps 4–8 performed additional ARX operations to mix (sequentially) the 1024 entries of P/Q tables. In particular, line 6 and line 7 are similar to its key generation output stage (next) except no output is produced. Therefore, it can be viewed as "burn-in" steps so that (hopefully) it will produce P/Q tables following the desirable uniformity via a series of non-linear operations.

4. Clearly, the initialization procedure is long because it would require long ARX operations to yield P/Q tables with desirable "uniformity" and "randomness" properties needed for

its key stream generation procedure. However, strong theoretical justification for such goals can not be achieved for any kind of Key/IV selected.

12.1.2 The Salsa Family and ChaCha

Both Salsa20 and its variant, ChaCha, operate on a vector of sixteen 32-bit words and produce a sequence of sixteen keystream of 32-bit words. They take as input eight words of key $\mathbf{k} = (k_0, k_1, ..., k_7)$, two words of nonce $\mathbf{v} = (v_0, v_1)$, four words of pre-defined constant $\mathbf{c} = (c_0, c_1, c_2, c_3)$, and two words of counter $\mathbf{t} = (t_0, t_1)$. The 16-dim state/input vector is commonly viewed as a 4×4 matrix and only the 64-bit counter $\mathbf{t}$ will be updated throughout the keystream generation stage with the same 64-bit nonce $\mathbf{v}$. For a given 4×4 state matrix, Salsa20 and ChaCha use various simple and effective ARX (Addition, Rotation, XOR) transformations in fixed rounds (say, R rounds) of iterations. The security of the ciphers is improved as the number of rounds, R, increases. It is common to use Salsa20/R and ChaCha$R$ to denote the specific R-round ciphers, and Salsa20 and ChaCha denote the family of ciphers. To propose and justify various simple and effective extensions on ChaCha, we describe more details about the original ciphers next.

Components and Operation of Salsa and ChaCha

Salsa20/R is a family of stream ciphers introduced in Bernstein (2008b) which is based on three simple operations: Addition, Rotation, XOR (ARX) applied in a fixed number of rounds, R. Bernstein (2008b) recommends Salsa20/20 for most applications. Salsa20/12 was submitted to the eSTREAM for Profile 1 (software-based ciphers) and was chosen as a finalist in the Phase 3 design (see, Robshaw and Billet 2008).

The ChaCha cipher, proposed by Bernstein (2008a), is closely related to the Salsa20 family and aims to increase the amount of diffusion without increasing complexity. Both Salsa20 and ChaCha stream ciphers are relatively simple in their design. Listed below are different components and operations between Salsa20 and ChaCha stream ciphers:

1. Choose the initial state vector of sixteen 32-bit words viewed as a 4×4 matrix for Salsa20 and ChaCha. Respectively, as

$$
\mathbf{X}_S^0 = \begin{pmatrix} c_0 & k_0 & k_1 & k_2 \\ k_3 & c_1 & v_0 & v_1 \\ t_0 & t_1 & c_2 & k_4 \\ k_5 & k_6 & k_7 & c_3 \end{pmatrix}, \quad \mathbf{X}_C^0 = \begin{pmatrix} c_0 & c_1 & c_2 & c_3 \\ k_0 & k_1 & k_2 & k_3 \\ k_4 & k_5 & k_6 & k_7 \\ t_0 & t_1 & v_0 & v_1 \end{pmatrix}, \tag{12.1}
$$

where

a. (c_0, c_1, c_2, c_3) is a predefined constant 64-bit vector with $c_0 = $ 0x61707865, $c_1 = $ 0x3320646e, $c_2 = $ 0x79622d32, $c_3 = $ 0x6b206574,

b. $(k_0, \dots, k_7)$ is the user key of 256 bits,

c. (v_0, v_1) is the nonce that can only be used once, and

d. (t_0, t_1) is the counter which will be updated sequentially throughout the keystream generation stage. The range of the counter $(t_0 + t_1 2^{32})$ is from 0 to $2^{64} - 1$.

According to Bernstein (2008a), the change of order in the state matrix has no effect on security but ChaCha is more efficient on some computing platforms.

2. Choose different quarter-round functions, $\mathbf{Q}(a, b, c, d)$ to perform an ARX transformation on four 32-bit words (a, b, c, d). For Salsa20

$$
\begin{aligned}
b &= b \oplus ((a + d) \lll 7) \\
c &= c \oplus ((b + a) \lll 9) \\
d &= d \oplus ((c + b) \lll 13) \\
a &= a \oplus ((d + c) \lll 18)
\end{aligned}
\tag{12.2}
$$

and for ChaCha

$$
\begin{aligned}
a &= a + b; \ d = d \oplus a; \ d = d \lll 16 \\
c &= c + d; \ b = b \oplus c; \ b = b \lll 12 \\
a &= a + b; \ d = d \oplus a; \ d = d \lll 8 \\
c &= c + d; \ b = b \oplus c; \ b = b \lll 7,
\end{aligned}
\tag{12.3}
$$

where, as defined previously in Sect. 1.1, $\oplus$ indicates XOR, $+$ means addition modulo 2^{32} and $x \lll r$ is r-bit rotation of a 32 bit word x by r bits to the left.

Salsa20 and ChaCha operate on only 4 words at a time to minimize memory access using an operation known as a quarter-round (Bernstein 2008b), Bernstein (2008a). In the ChaCha quarter-round, unlike the Salsa20, each input word affects each output word and diffusion occurs more quickly through the ChaCha quarter-round. According to Bernstein (2008a), a ChaCha quarter-round changes on average 12.5 output bits compared to 8 output bits in a Salsa20 quarter-round. This increased diffusion improves the quality of the stream, and hence makes ChaCha stronger than Salsa20.

3. A general full-round function (systematically) partitions the 4×4 state matrix into four vectors of four words, then uses the given quarter-round function (sequentially or in parallel) on the four vectors. To describe the partition, let the current state matrix be

$$
\mathbf{X} = \begin{pmatrix}
x_0 & x_1 & x_2 & x_3 \\
x_4 & x_5 & x_6 & x_7 \\
x_8 & x_9 & x_{10} & x_{11} \\
x_{12} & x_{13} & x_{14} & x_{15}
\end{pmatrix}
\tag{12.4}
$$

and define a "partition" of $\mathbf{X}$

$$
P(\mathbf{X}) = (\mathbf{x}_0, \mathbf{x}_1, \mathbf{x}_2, \mathbf{x}_3).
\tag{12.5}
$$

Salsa20 and ChaCha have different choices of partitions and they also use different partitions for their odd-round and even-round transformations. We believe that specific choices of the partition for Salsa20 and ChaCha were carefully made to maximize security and efficiency.

a. In odd number rounds, both Salsa20 and ChaCha used the same term *column-rounds* with slightly different partitions:

$$
\text{(Salsa20 : Odd round)} \quad
\begin{aligned}
\mathbf{x}_0 &= (x_0, \ x_4, \ x_8, \ x_{12}), \\
\mathbf{x}_1 &= (x_5, \ x_9, \ x_{13}, \ x_1), \\
\mathbf{x}_2 &= (x_{10}, \ x_{14}, \ x_2, \ x_6), \\
\mathbf{x}_3 &= (x_{15}, \ x_3, \ x_7, \ x_{11}),
\end{aligned}
\qquad (12.6)
$$

and

$$
\text{(ChaCha : Odd round)} \quad
\begin{aligned}
\mathbf{x}_0 &= (x_0, \ x_4, \ x_8, \ x_{12}), \\
\mathbf{x}_1 &= (x_1, \ x_5, \ x_9, \ x_{13}), \\
\mathbf{x}_2 &= (x_2, \ x_6, \ x_{10}, \ x_{14}), \\
\mathbf{x}_3 &= (x_3, \ x_7, \ x_{11}, \ x_{15}).
\end{aligned}
\qquad (12.7)
$$

For ChaCha, $\mathbf{x}_i$ corresponds to the columns of $\mathbf{X}$ in Eq. 12.4 and for Salsa20, $\mathbf{x}_i$ corresponds to the (rotated) columns of $\mathbf{X}$.

b. In even number rounds, Salsa20 used *row-rounds* while ChaCha used *diagonal-rounds*:

$$
\text{(Salsa20 : Even round)} \quad
\begin{aligned}
\mathbf{x}_0 &= (x_0, \ x_1, \ x_2, \ x_3), \\
\mathbf{x}_1 &= (x_5, \ x_6, \ x_7, \ x_4), \\
\mathbf{x}_2 &= (x_{10}, \ x_{11}, \ x_8, \ x_9), \\
\mathbf{x}_3 &= (x_{15}, \ x_{12}, \ x_{13}, \ x_{14}),
\end{aligned}
\qquad (12.8)
$$

and

$$
\text{(ChaCha : Even round)} \quad
\begin{aligned}
\mathbf{x}_0 &= (x_0, \ x_5, \ x_{10}, \ x_{15}), \\
\mathbf{x}_1 &= (x_1, \ x_6, \ x_{11}, \ x_{12}), \\
\mathbf{x}_2 &= (x_2, \ x_7, \ x_8, \ x_{13}), \\
\mathbf{x}_3 &= (x_3, \ x_4, \ x_9, \ x_{14}).
\end{aligned}
\qquad (12.9)
$$

For Salsa20, $\mathbf{x}_i$ corresponds to the (rotated) rows of $\mathbf{X}$ in Eq. 12.4 and for ChaCha, $\mathbf{x}_i$ corresponds to the (rotated) "diagonal" of $\mathbf{X}$ which turns out to be the "transpose" of the Salsa20's odd-round (its column round).

For each specified partition in Eqs. 12.6, 12.7, 12.8, and 12.9, we perform a full-round transformation with four quarter-round transformations: $\mathbf{Q}(\mathbf{x}_0), \mathbf{Q}(\mathbf{x}_1), \mathbf{Q}(\mathbf{x}_2)$ and $\mathbf{Q}(\mathbf{x}_3)$. This process is repeated $R/2$ times (with odd round and then even round) so that it has a total of R rounds of full-round transformations (or $4R$ quarter-round transformations). For Salsa20, we let

$$\mathbf{X}_S^R = \mathbf{H}_S(\mathbf{X}_S^0; R), \tag{12.10}$$

and for ChaCha, we let

$$\mathbf{X}_C^R = \mathbf{H}_C(\mathbf{X}_C^0; R). \tag{12.11}$$

General Procedure of Salsa and ChaCha

To simplify the notation, we drop the subscript S (for Salsa20) and C (for ChaCha) as described above. Specifically, we use $\mathbf{X}^0$ to denote the initial state matrix either $\mathbf{X}_S^0$ or $\mathbf{X}_C^0$. Likewise, we use $\mathbf{X}^R$ to denote the transformed matrix either $\mathbf{X}_S^R$ or $\mathbf{X}_C^R$, we use $\mathbf{H}$ to denote either $\mathbf{H}_S$ or $\mathbf{H}_C$.

With the main components and operations specified, we can then formally describe the general common procedure of Salsa20 and ChaCha as in Algorithm 12.3. The common general procedure for Salsa20 and ChaCha can be shown in Fig. 12.1. Bernstein (2008a) provided timing comparisons, using the official eSTREAM benchmarking framework, for ChaCha and the fastest implementation of Salsa20 and found that the comparison results are highly platform-dependent.

Algorithm 12.3: General procedure of Salsa and ChaCha

Step 0: Create an initial state matrix $\mathbf{X}^0$ as in Eq. 12.1. Let i be the running index of the iteration. Counter (t_0, t_1) can be set as $i = i_0$, where i_0 can be chosen as any number. For simplicity, we set $i = 0, t_1 = 0$, and initial "carry" $c = 0$.

Step 1: At the i-th stage, initialize state matrix $\mathbf{X}^0$ in Eq. 12.1 with counter $t_0 = i, t_1 = t_1 + c$.

Step 2: Compute R-round transformation as $\mathbf{X}^R = \mathbf{H}(\mathbf{X}^0; R)$ in Eq. 12.10 or Eq. 12.11.

Step 3: Output sixteen 32-bit words

$$\mathbf{Z}i = \mathbf{X}^0 \boxplus \mathbf{X}^R.$$

Increase index $i = (i + 1)\&\mathbf{1}_{32}$, set the carry bit (when $i = 0 \mod 2^{32}$) and repeat the above Step 1 through Step 3.

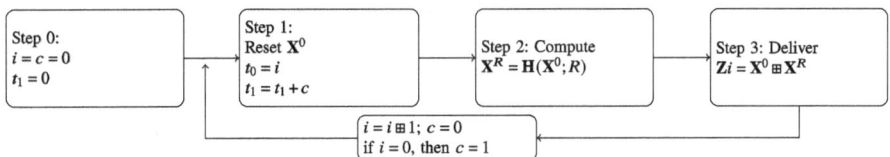

Fig. 12.1 Diagram of ChaCha

12.1.3 Rabbit

Rabbit cipher, proposed by Boesgaard et al. (2003), has an input of a 128-bit key, an optional 64-bit initialization vector. Internally, the cipher uses a state vector and a counter vector for each of eight 32-bit words and an additional one-bit for storing "carry". The output of the cipher was then produced from the eight updated state vectors into four 32-bit words per iteration. The cipher's strength rests on a strong mixing of its inner state and counter vectors during successive iterations. The mixing function was based on arithmetical operations that are available on a modern processor without the need for S-boxes or lookup tables. The mixing function used a g-function with a squaring function, and the ARX operations which consists of logical XOR, bit-wise rotation with fixed rotation amounts, and addition modulo 2^{32}.

Rabbit cipher can be divided into several distinct modules. The first module was to initialize the internal counter and state vectors with some expansion scheme with input key and optional IV vectors to initialize the internal state and counter variables with a series of non-linear transformations to achieve random and uniform distribution for these internal vectors. The second module was to update the counter variables using some counter-based scheme and carry bit so that a long period length of the counter system can yield a long period length of 2^{256}. The third module used a non-linear function, g-function. to mix the state vector and counter vector to produce a new G vectors which was then used to update the state vector with another set of eight non-linear transformations. Finally, Rabbit produces four 32-bit words from the state vector of eight 32-bit words from various combinations of sub-words of eight state variables.

Rabbit stream cipher is recognized as an excellent cipher which has been known to enjoy strong security and great generating efficiency. In this paper, we propose some simple and effective "enhancement" to Rabbit on previously mentioned modules to improve some possible "problems" identified. We will start the proposed enhancement in the reverse order of the Rabbit modules: (a) update state vector with g function (b) update counter vector (c) initialization of state/counter variables.

According to its designer, $g(x)$ was a key component in the Rabbit design squaring a 32-bit number to produce a 64-bit number, and then combining the left half and the right half of that square number with xor, to produce a 32-bit result. As pointed out first by Aumasson (2007), however, such a function may produce a small bias in the output of Rabbit and result in a distinguisher with a very large number of runs. Therefore, it is not considered a minor threat to Rabbit's security because its complexity is significantly higher than the brute-force of the key space. We will propose an alternative function to replace the g-function.

Another possible problem for Rabbit is its counter-updating scheme which was based on a simple counter-updating scheme. Such a counter-based scheme can ensure an extremely long period length of the count system with its period length reaching $2^{256} - 1$. Such an updating scheme can be viewed as a simple way to produce a sequence of random numbers with constant increments (with possible carry added) at each iteration. There are many good random number generators that one can choose with a great distributional property and huge period length. We propose to replace this counter-update scheme with a set of eight good external generators.

Finally, Rabbit's initialization scheme used some key/IV expansion scheme to fill in the state/counter vectors with an additional series of non-linear ARX operations on these vectors. However, there is no justification that Rabbit's initialization scheme will ensure a nice distribution property even after a series of non-linear ARX operations with any choice of key/IV. A better strategy is us an additional good external generator and Key/IV are used initialized their seed vectors. The external generators then can be used to fill the contents of state and counter variables. Typical good external generators will not be affected by the choice of initial seeds and key/IV information will be quickly "diluted" after a short iterations of "burn-in" period. This procedure of using external generator to fill in initial tables/vectors was proposed first in Deng et al. (2018).

The description of `Rabbit` stream cipher is proposed in Boesgaard et al. (2003). To describe `Rabbit`, we used the following for 32-bit integers x, y.

1. $x \gg n$ denotes the right shift operator of x, $x \ll n$ denotes the left shift operator of x
2. $x \| y$ dnotes the concatenation of the bit strings of x, y, $x \oplus y$ is the bit-wise logical XOR of x and y, $x \& y$ is the bit-wise logical AND of x and y, $x + y$ means $x + y \mod 2^{32}$.
3. $\lfloor x/y \rfloor$ denotes the floor/integer function of a value x/y.
4. $x \ggg n$ denote the right rotation operator, $((x \gg n) \oplus (x \ll (32 - n)))$ and $x \lll n$ denote the left rotation operator $((x \ll n) \oplus (x \gg (32 - n)))$.

There are several components for `Rabbit` cipher which can divided into three categories: (1) input with key/IV (2) internal state/counter/carry variables and (3) output variables as in Algorithm 12.4.

In total, there are 513 "internal" bits with eight 32-bit state variables, eight 32-bit counters, and one counter carry bit. Generally speaking, a series of non-linear transformations on these variables to update variables. The functions used are typical operations, known as ARX (Addition, Rotation, and XOR) operations, which are useful to make non-linear transformations (mostly with rotation) and mixing/addition can hide the individual values and also improve the distributional property as "combination" generator. The "goal" is to "defuse" the information about **K** and **IV** while the state variables **X**. A high-level description can be described as follows:

1. **Key/IV initialization.** These internal state variables **X** and counter variables $\mathbb{C}$ will be "initialized" with complicated functions of **K** and (optional) **IV**. Use a series of non-linear transformations to "defuse" the information about **K** and **IV** while the state variables **X** are "random" following a "uniform" distribution.

Algorithm 12.4: Rabbit cipher

---

1. input variables: **K** and **IV**
   a. **K** is an input secret key of 128 bits, it can be divided into 8 subkeys of 16-bit each

   $$\mathbf{K} = [k_7, k_6, k_5, k_4, k_3, k_2, k_1, k_0],$$

   b. **IV** is an optional input secret key of 64 bits, it can be divided into 2 or 4 subkeys of equal size each

   $$\mathbf{IV} = [v_3, v_2, v_1, v_0],$$

2. internal variables: **X** and **C**
   a. **X** is an internal state vector of size 8 words,

   $$\mathbf{X} = [x_7, x_6, x_5, x_4, x_3, x_2, x_1, x_0]$$

   it can be further divided into 2 subkeys of 16-bit each from each word

   $$x_j = [x_{j1} \| x_{j0}], \quad j = 0, \cdots, 7.$$

   b. **C** is an internal "counter" vector of size 8 words,

   $$\mathbf{C} = [c_7, c_6, c_5, c_4, c_3, c_2, c_1, c_0]$$

   c. ϕ^* is an internal "carry" bit of size one, which is storing the "carry" from the most recent calculation/update of "counter" word.

3. output variables: **S**
   a. The output variables have four 32-bit words computed from internal states variables of 8 words, **X**
   b.

   $$\mathbf{S} = [s_3, s_2, s_1, s_0].$$

---

2. **Counter vector update.** Counter variables $\mathbb{C}$ will be updated with a sequence of equations of a counter-based scheme with extra "carry" to be added in sequential update/computation.

3. **State vector update.** Produce a G-vector of eight words using a specialized non-linear transformation on $g(x)$ function to mix the current state vector and updated counter vectors. Update state vector, **X**, with another set of transformations on the G vector.

4. **Output extraction scheme.** Choose four extract functions for four output variables, SS, from the updated eight state variables, **X**.

5. Repeat (2), (3), (4) for next outputs.

The details of `Rabbit` design are as follows:

Key/IV Setup Initialization Scheme

Let $\mathbf{K}$ be an input secret key of 128 bits with 8 subkeys of 16-bit each

$$\mathbf{K} = [k_7, k_6, k_5, k_4, k_3, k_2, k_1, k_0].$$

Key Setup

Using the subkeys of $\mathbf{K}$, initialize the state variables and counter variables as

$$\mathbf{X} = [k_3\|k_4, k_7\|k_6, k_2\|k_1, k_5\|k_4, k_0\|k_7, k_3\|k_2, k_6\|k_5, k_1\|k_0]$$

and

$$\mathbf{C} = [k_7\|k_0, k_2\|k_3, k_5\|k_6, k_0\|k_1, k_3\|k_4, k_6\|k_7, k_1\|k_2, k_4\|k_5]$$

To diffuse the key/IV information and randomize $\mathbf{X}$ and $\mathbf{C}$ with desirable uniformity property, it is recommended to use the next-state function (to be formally defined next) in four iterations. Its main purpose is to mix and then update $\mathbf{X}$ and $\mathbf{C}$ via a series of ARX operations to diminish correlations between bits in the key and bits in the internal state variables to achieve better "randomness" and "uniform distribution" properties. Finally, to prevent possible recovery of the key by inversion of the counter system, the counter variables are re-initialized according to:

$$\mathbf{C} = \mathbf{C} \oplus \text{Rotate}(\mathbf{X}, 4) = \mathbf{C} \oplus [x_3, x_2, x_1, x_0, x_7, x_6, x_5, x_4].$$

IV Setup

View $\mathbf{IV}$, the optional input key of 64 bits, as 4 subkeys of equal size (16-bit) each:

$$\mathbf{IV} = [v_{11}, v_{10}, v_{01}, v_{00}]$$

From these four subkeys, create 8 32-bit words as

$$\mathbf{V} = [v_{10}\|v_{00}, v_{11}\|v_{10}, v_{11}\|v_{01}, v_{01}\|v_{00}, v_{10}\|v_{00}, v_{11}\|v_{10}, v_{11}\|v_{01}, v_{01}\|v_{00}]$$

and then mix (via $\oplus$) with the eight counter variables as

$$\mathbf{C} = \mathbf{C} \oplus \mathbf{V}$$

As before, iterate four times with the next-state function to further mix the state and counter variables.

Next-State Function

The next-state function is the most important part of $\texttt{Rabbit}$ cipher. It has three steps:

1. counters are updated sequentially with carry bit computed for next counter update
2. g-values are computed with the non-linear function of old state-variable and updated counter-variable.
3. state variables are updated from the newly computed g-values.

For better modularity, the implementation of these steps can be thought of as calling three different functions.

Counter Update

First, eight constants are needed in the scheme updating eight counters:

$$
\begin{aligned}
\alpha_0 &= 0x4D34D34D & \alpha_1 &= 0xD34D34D3 \\
\alpha_2 &= 0x34D34D34 & \alpha_3 &= 0x4D34D34D \\
\alpha_4 &= 0xD34D34D4 & \alpha_5 &= 0x34D34D34 \\
\alpha_6 &= 0x4D34D34D & \alpha_7 &= 0xD34D34D3
\end{aligned}
\tag{12.12}
$$

They are used to update the counter variables in addition to a "carry" from the previous counter update.

Let ϕ^* be the indicator for the carry from the previous counter variable update during the iteration of all eight counter variables. We can set $\phi^* = 0$. For $j = 0, 1, 2, \ldots, 7$, update the j-counter as

$$
\begin{aligned}
R &= c_j + \alpha_j + \phi^*; \\
c_j^* &= R \quad \mathrm{mod}\ 2^{32}; \\
\phi^* &= I(R \geq 2^{32}),
\end{aligned}
\tag{12.13}
$$

where $\phi^* = I(A)$ is an indicator function of a set A. After the update sequence above, we have updated $\mathbf{C}$ as

$$
\mathbf{C}^* = [c_7^*, c_6^*, c_5^*, c_4^*, c_3^*, c_2^*, c_1^*, c_0^*].
$$

Non-linear G-Function: G-Values

Then the non-linear g-function below is used to update $\mathbf{G} = [g_7, g_6, g_5, g_4, g_3, g_2, g_1, g_0]$ as

$$
\mathbf{G} = G(\mathbf{X}, \mathbf{C}^*) = g(\mathbf{X} + \mathbf{C}^*) = ((\mathbf{X} + \mathbf{C}^*)^2 \oplus (\mathbf{X} + \mathbf{C}^*)^2 \gg 32) \quad \mathrm{mod}\ 2^{32}, \tag{12.14}
$$

where for vector x, x^2 is a component-wise square.

State Update

After $\mathbf{G}$ has been computed, the next state variables, $\mathbf{X}$, are computed/updated from a subset of $\mathbf{G}$ with some of their substrings. Two functions will be used for the state update functions on 3 integers x, y, z.

1. $H_1(x, y, z; a, b) = x + (y \ll a) + (z \ll b) \mod 2^{32}$.
2. $H_2(x, y, z; a) = x + (y \ll a) + x \mod 2^{32}$.

$$
\begin{aligned}
x_0^* &= H_1(g_0, g_7, g_6; 16, 16) & x_1^* &= H_2(g_1, g_0, g_7; 8) \\
x_2^* &= H_1(g_2, g_1, g_0; 16, 16) & x_3^* &= H_2(g_3, g_2, g_1; 8) \\
x_4^* &= H_1(g_4, g_3, g_2; 16, 16) & x_5^* &= H_2(g_5, g_4, g_3; 8) \\
x_6^* &= H_1(g_6, g_5, g_4; 16, 16) & x_7^* &= H_2(g_7, g_6, g_5; 8)
\end{aligned}
\tag{12.15}
$$

Update the counter and state variable

$$
\mathbf{C} \leftarrow \mathbf{C}^*, \quad \mathbf{X} \leftarrow \mathbf{X}^*.
$$

Output Extraction Scheme

The output variables, $SS = [s_3 \| s_2 \| s_1 \| s_0]$, have four 32-bit words computed from internal states variables of 8 words

$$
\mathbf{X} = [x_7, x_6, x_5, x_4, x_3, x_2, x_1, x_0]
$$

that can be further divided into 2 subkeys of 16-bit each as,

$$
x_j = [x_{j1} \| x_{j0}], \quad j = 0, \ldots, 7.
$$

Four 32-bit outputs are produced with some simple bit-wise combination of bytes from eight 32-bit outputs (updated) state vectors $\mathbf{X}$ as

$$
\begin{aligned}
s_0 &= (x_{00} \oplus x_{51}) \| (x_{01} \oplus x_{30}) \\
s_1 &= (x_{20} \oplus x_{71}) \| (x_{21} \oplus x_{50}) \\
s_2 &= (x_{40} \oplus x_{11}) \| (x_{41} \oplus x_{70}) \\
s_3 &= (x_{60} \oplus x_{31}) \| (x_{61} \oplus x_{10}).
\end{aligned}
\tag{12.16}
$$

12.1.4 Sosemanuk

Sosemanuk, proposed by Berbain et al. (2008), is another synchronous software-oriented stream cipher with its key length variable between 128 and 256 bits and a 128-bit initial value. Any key length is claimed to achieve 128-bit security. The Sosemanuk, along with

HC-128, Rabbit, and Salsa20/12, is one of the final four Profile 1 (software) ciphers selected for the STREAM Portfolio.

According to Mukherjee (2013), Sosemanuk is basically not very popular due to its complicated structure with several attacks already demonstrated and some of them may pose a real threat too. Therefore, we will not discuss it further here.

12.2   Profile 2: Hardware Ciphers

There are three finalists in Profile 2: (a) Grain v1 (b) MICKEY 2.0 (c) Trivium selected in hardware ciphers. These designs limited hardware environments where gate count, power consumption, and memory. We Briefly describe them next.

12.2.1 Grain V1

Grain v1 was proposed by Hell et al. (2007) which is based on two shift registers and a nonlinear filter function. According to its designers, the cipher has the additional feature that the speed can be increased at the expense of extra hardware. The key size is 80 bits and a 64-bit IV. No attack faster than an exhaustive key search has been identified.

Grain v1 has a 160-bit internal state consisting of an 80-bit linear feedback shift register (LFSR) and an 80-bit non-linear feedback shift register (NLFSR). Grain v1 updates one bit of LFSR and one bit of NLFSR state for every bit of ciphertext released by a nonlinear filter function. The 80-bit NLFSR is updated with a nonlinear 5-to-1 Boolean function and a 1-bit linear input selected from the LFSR. The nonlinear 5-to-1 function takes as input 5 bits of the NLFSR state. The 80-bit LFSR is updated with a 6-to-1 linear function. During keying operations the output of the cipher is additionally feedback as linear inputs into both the NLFSR and LFSR update functions.

Grain v1 is simple and is an attractive choice for cryptanalysts and implementors alike with two shift registers: one with linear feedback and the second with nonlinear feedback; being the essential feature of the algorithm family. These registers, and the bits that are output, are coupled by means of very lightweight, but judiciously-chosen boolean functions.

12.2.2 MICKEY 2.0

MICKEY 2.0 (which stands for Mutual Irregular Clocking KEYstream generator), proposed by Babbage and Dodd (2008), is aimed at resource-constrained hardware platforms. According to its designers, it is intended to have low complexity in hardware, while providing a high level of security. It uses irregular clocking of shift registers, with some novel techniques to balance the need for guarantees on period and pseudorandomness against the need to avoid certain cryptanalytic attacks.

MICKEY 2.0 has a 80-bit key and an initialization vector with up to 80 bits in length which has secret state consists of two 100-bit shift registers, one linear and one nonlinear, each of which is irregularly clocked under control of the other. According ot its designers, MICKEY 2.0 has specific clocking mechanisms contribute cryptographic strength while still providing guarantees on period and pseudorandomness. The designers have also specified a scaled-up version of the cipher called MICKEY-128 2.0, which takes a 128-bit key and an IV up to 128 bits.

12.2.3 Trivium

Trivium is a hardware-efficient (profile 2), synchronous stream cipher proposed by De Canniere and Preneel (2008) with an 80-bit key and 80-bit initialization vector (IV). Trivium has a secret state of 288 bits with three interconnected nonlinear feedback shift registers of length 93, 84, and 111 bits, respectively. The cipher operation consists of two phases: the key and IV set-up and the keystream generation.

1. Initialization is very similar to keystream generation and requires 1152 steps of the clocking procedure of Trivium.
2. The keystream is generated by repeatedly clocking the cipher, wherein each clock cycles three state bits are updated using a non-linear feedback function, and one bit of keystream is produced and output.

The cipher specification states that 264 keystream bits can be generated from each key/IV pair.

Methods to Improve Security of the Linear Generators

This chapter discusses various techniques to enhance the performance of random number generators. One improved generator over a basic one presented in Sect. 13.1, the combination generator, proposed first by Wichmann and Hill (1982), is constructed by taking the fractional part of the sum of several random number generators. It is considered one of the most popular random number generators. Its empirical performance is superior to the classical Lehmer congruential generator. Specifically, it can be proved that the combination generator method is superior to each component random number generator method, in terms of (1) uniformity and (2) independence.

There are several different methods to improve generators, such as shuffle methods in Sect. 13.2.1. The TAC method involves twisting and combining two random number generators by multiplying constants and taking the modulus. The twisting operation involves multiplying a random variate by a large constant to twist its bits, while the combining operation adds two random number generators together.

13.1 Addition, Rotation, XOR (ARX) and Its Variants

ARX transformation

The ARX transformation operates at the bit level, using three basic operations: addition (A), rotation (R), and exclusive OR (XOR). These operations are performed sequentially or in a specific order to create a complex transformation. ARX transformations are known for their simplicity and efficiency, as they can be implemented using basic logical and arithmetic operations that are typically available in modern computer architectures.

© The Author(s), under exclusive license to Springer Nature Switzerland AG 2025
L. Deng et al., *Random Number Generators for Computer Simulation and Cyber Security*, Synthesis Lectures on Mathematics & Statistics,
https://doi.org/10.1007/978-3-031-76722-7_13

- Addition (A): In this operation, the binary values of two operands are added together using modular addition (usually modulo 2^n, where n is the word size). This operation is also sometimes called modular addition or bitwise addition.
- Rotation (R): Rotation involves shifting the bits of a binary value to the left or right. The rotated bits wrap around, so the bits that are shifted off one end are reinserted at the other end. Left rotation is denoted as ROL (Rotate Left), and right rotation is denoted as ROR (Rotate Right).
- Exclusive OR (XOR): The XOR operation takes two binary values and compares their bits. If the corresponding bits are different, the result is 1; otherwise, it is 0. XOR is commonly used in cryptography for its properties, such as its ability to introduce confusion and diffusion in the data.

Consider the three examples below, assuming that the system works with 8-bit numbers, x is 00111011, and y is 11100010.

Example 13.1 (*Addition of x and y*)

If a bit-wise addition is operated on x and y, the bit-wise operation can be shown in the following standard Algorithm:

$$
\begin{array}{r}
00111011 \\
\&\ \ 11100010 \\
\hline
1\ 00111101
\end{array}
$$

The result of the bit-wise addition $x \& y$ is "1" 00111101. Note that a carry is generated in the most significant bit, represented by the carry-over digit "1" at the beginning.

Example 13.2 (*Exclusive OR of x and y*)

$x \oplus y$ performs the XOR operation between x and y bit by bit, which can be shown as the following standard algorithm.

$$
\begin{array}{r}
00111011 \\
\oplus\ \ 11100010 \\
\hline
11011001
\end{array}
$$

Example 13.3 (*Left Rotation of x by 3 bits*)

To compute $x \lll 3$, we can utilize the formula $(x \ll 3) \oplus (x \gg (8-3))$. First, let's determine $(x \ll 3)$, which can be obtained by shifting x to the left by three positions. This results in 11011000 by appending three zeros to the right and discarding the top three bits. Similarly, $(x \gg (8 - 3))$ equals 00000001, obtained by shifting x to the right by 5 positions. Finally, we can perform the bit-wise XOR ($\oplus$) operation between $(x \ll 3)$ and $(x \gg (8 - 3))$ as follows:

$$
\begin{array}{r}
11011000 \\
\oplus\; 00000001 \\
\hline
11011001
\end{array}
$$

By combining these three operations in different ways and applying them repeatedly, the ARX transformation creates a complex and nonlinear mapping between the input data and the output data. This mapping is designed to provide confusion and diffusion properties, making it difficult for an attacker to analyze and reverse-engineer the encryption algorithm.

Definition 13.1 Let's define the operations in terms of binary manipulation for two integers, denoted as x and y:

- $x \& y$ represents the result obtained by performing bit-wise addition on their binary representations,
- $x \gg d$ denotes the operation of logical right shifting x by d bits in its binary representation, and $x \ll d$ denotes the operation of logical left shifting x by d bits in its binary representation,
- $x \oplus y$ corresponds to the bit-wise "exclusive or" operation between x and y, that is, If the bits are different (one bit is 0 and the other is 1), the result is 1. If the bits are the same (both 0 or both 1), the result is 0.
- $x \lll d$ corresponds to rotating x by d bits, defined as the result of $(x \ll d) \oplus (x \gg (b - d))$, where b represents the maximum number of bits in the system.

Non-linear transformations for the mixing function

We will now focus on selecting the "transformation" functions, e.g., $f_j(\cdot)$, $j = 1, 2, 3, 4$ in. There are numerous potential choices for these functions, and we suggest using bit rotations similar to those in the study by Wu (2004). The bit rotation function for a 32-bit variate x is of the form $g_r(x) = (x \gg r) \oplus (x \ll (32 - r))$, where r is the size of bit rotation. For any combination of $1 \le r_1, r_2, r_3, r_4 \le 31$, we can define

$$f_j(x) = g_{r_j}(x) = (x \gg r_j) \oplus (x \ll (32 - r_j)), \quad j = 1, 2, 3, 4. \tag{13.1}$$

One benefit of using bit rotation functions in conjunction with the efficient $\boxplus$ function to obtain the output U_i is to have a large number of possible design parameter combinations.

Many of the efficient baseline generators considered for the internal states X and Y, such as the DX-k generators, only generate 31 bits at each iteration so that the leading bit of the

32-bit word is always zero. If both baseline generators are 31-bit generators, then all the internal and intermediate states also have 31 bits in length. While simple (no transformation) addition of the three 31-bit variates $(X \boxplus Y \boxplus W)$ may still produce a "carry" to the leading bit, the leading bit would not behave like the desired uniform distribution over 0/1. We can solve this problem through rotations. By rotating the three variates with different sizes, their leading bits are rotated to different locations; and the subsequent summation of the rotated variates, $f_1(X) \boxplus f_2(Y) \boxplus f_4(W)$, helps "even out" the distribution over the entire 32 bits. This gives another benefit of using rotation functions in the output function.

As stated before, one class of generators produces a sequence of numbers based on a linear recurrence relation. It is called the lagged Fibonacci generator named after the Fibonacci sequence. The lagged Fibonacci generator is defined as follows,

$$X_n = (X_{n-q} \pm X_{n-p}) \bmod m \tag{13.2}$$

In this equation, X_n represents the nth random number in the sequence, X_{n-q} and X_{n-p} are the q and pth previous numbers in the sequence, respectively, and m is a modulus value that determines the range of the generated numbers.

Another type of generator studied in Ferrenberg et al. (1992) is based on "subtract-with-borrow (carry)" and "add-with-carry" methods which are proposed by Marsaglia and Zaman (1991). They are simple modifications of the Lagged Fibonacci Generator defined by including a carry from previous subtraction or addition operation:

$$x_n = x_{n-q} - x_{n-p} - c_{n-1} \quad \bmod m,$$

where c_n has the value 1 if $x_{n-q} - x_{n-p} - c_{n-1} < 0$; otherwise 0. Such generators are referred to as SWB generators or SWC generators in general. We will use the term SWC as in Ferrenberg et al. (1992). The Add-with-carry (AWC) generators can be similarly defined, though they will not be further discussed here.

The choice of p, q, m determines the period of the SWC generators. As shown in Marsaglia and Zaman (1991), the period of SWC$(p, q; m)$ generator is $\phi(m^p - m^q + 1)$ (see also Crandall and Pomerance 2006, Theorem 8.2.6, p. 367), where $\phi(x)$ is the Euler "totient" function of x, which is the number of integers between 1 and x that are relatively prime to x. Hence, if $M = m^p - m^q + 1$ is a prime number, then the generator has a maximum period of $m^p - m^q$.

Generally speaking, SWC generators can be very efficient because no multiplication is required. In particular, the SWC(43, 22; $2^{32} - 5$) generator, is called SWC in Ferrenberg et al. (1992), and its period is $10^{414.2}$. This kind of generator is slightly more complicated than the LFG and it is almost as efficient. However, similar problems in its empirical performance is also reported in Ferrenberg et al. (1992).

13.2   Shuffling and Its Variants

13.2.1 Shuffle Methods

Chapter 2 highlights that linear generators such as MRGs and LCGs lack security due to their inherent linearity, which allows for the computation of their next value by solving linear recurrence equations based on previously observed values (Lidl and Niederreiter 1994). Consequently, these classical generators are unsuitable for computer security applications. To address the linearity issue, MacLaren and Marsaglia (1965), also known as Algorithm M – Randomizing by shuffling (Knuth 1998, introduced the shuffle method), as shown in Fig. 13.1.

This method involves incorporating an auxiliary variable Y to "shuffle" the output sequence of X using an auxiliary table $(T[\cdot])$ that stores the values of X_i. At each iteration, the current value of the auxiliary generator Y_i and an index-selection function $r(y)$ are used to select the $r(Y_i)$th entry in the auxiliary table as the next output $V_i \equiv T[r(Y_i)]$. Subsequently, the slot is refilled with the current value X_i of the baseline generator: $T[r(Y_i)] \leftarrow X_i$.

The shuffle generator offers advantages such as enhancing the period length and empirical performance of the baseline generator, as well as breaking the linear lattice structure of the baseline generator. Consequently, predicting the next value becomes challenging unless the original order of the generated variates can be restored. However, from a security perspective, this generator still has certain weaknesses. Firstly, while the output order is shuffled, all the outputs produced by the baseline generator X are revealed. Secondly, the auxiliary generator Y is not fully utilized to further improve the security of the shuffle generator. These weaknesses make it easy to predict future values, particularly when an LCG is employed as the baseline generator. These observations inspired the proposal of a more comprehensive shuffling method that utilizes two generators, with each serving as a baseline generator and an auxiliary generator to the other baseline generator.

Bays and Durham Algorithm
Algorithm B in Knuth (1998) are as in Fig. 13.2.

1. Its output used as next input variate ("feedback") as the auxiliary PRNG (self-shuffling).
2. No need to have additional PRNG.
3. Period length is unknown but increases with table size.

Fig. 13.1 Diagram of shuffle method

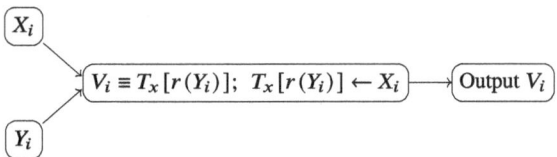

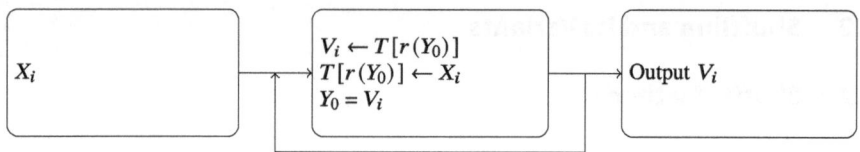

Fig. 13.2 Diagram of Bays and Durham

The shuffle generator has several advantages:

1. it will increase the period length from both generators
2. it will improve the empirical performance
3. it can break the linearity.

The shuffle generator still has some weakness:

1. its output variates are scrambled sequence of the *internal states* (X_i)
2. Internal state variables X_i's are *"unprotected"*: possible to reconstruct the original internal states (especially for LCG).
3. Auxiliary generator (Y_i) is *not fully utilized* to further improve the security of the generator.

13.2.2 Mutual-Shuffling Method

The shuffling method can break the linearity structure of a linear generator. Deng et al. (2018) proposed to extend a shuffle method to a mutual shuffle method with X and Y shuffling each other.

Let X and Y be two generators that serve as the baseline generators defined as:

$$X_i = f(X_{i-1}, \ldots, X_{i-k}) \mod p, \text{ and } Y_i = f(Y_{i-1}, \ldots, Y_{i-h}) \mod p, \quad i \geq 0,$$
$$(13.3)$$

We need two auxiliary tables (of the same size 2^C), $T_x[\cdot]$ and $T_y[\cdot]$, to store (as a buffer) the variates generated from X and Y, respectively. While a larger shuffle table tends to have a better scrambling property, we recommend its size to be between 32 and 512 for efficiency.

During the generation stage, we need an efficient entry-index-selection function, $r(z)$, to retrieve and update a shuffle table entry by selecting C bits of the generated value z. Here for simplicity, we choose $r(z) = z \& (2^C - 1)$, i.e., the least significant C bits of z. At the ith iteration, X_i and Y_i are generated by Eq. 13.3, respectively; and the output U_i can be produced by the mutual-shuffling method as illustrated in the diagram, Fig. 13.3.

The values (X_i, Y_i) and (V_i, W_i) are referred to as the values of the current "internal states" and "intermediate states" of the ith iteration, respectively. The final "output" is

SAFE: Mutual-Shuffling Method

14

In many areas of scientific research, random number generators (RNGs) play a very important role. For example, RNGs are used extensively in Monte Carlo integration, computer modeling, and simulations. In recent years, good and secure RNGs are becoming essential in many security-related applications such as the key stream generation for the one-time pad, automatic password generation, online gambling, and financial transactions. It is generally agreed that the RNG has been one of the weak links in various security applications. To design good RNGs suitable for both security and computer simulation applications is a challenging task.

For many computer simulation applications, it is common to use pseudo-random number generators (PRNGs) to produce a sequence of variates that is very hard to distinguish from a sequence of (truly) random numbers. An ideal PRNG should satisfy the property of HELP, which stands for high-dimensional equi-distribution, efficiency, long period-length, and portability. Recently, the Multiple Recursive Generator (MRG), an extension of the Linear Congruential Generator (LCG) (proposed by Lehmer 1951), has become very popular at least in the area of computer simulation. An MRG of order k computes its next variate via a linear combination of past k variates over a finite field of a large prime modulus m. In this sense, the Linear Feedback Shift Register (LFSR), another popular type of PRNGs, can be considered as a special case of the MRG with modulus 2. There have been great advances in the search of efficient maximum-period MRGs of a very large order. For the prime modulus $m = 2^{31} - 1$, the largest order of the maximum-period MRGs found so far is $k = 20897$. Deng et al. (2012b), which corresponds to a period length (i.e., $m^k - 1$) about $10^{195009.3}$. In addition to being efficient and portable, these generators have the nice property of equi-distribution over dimensions up to $k = 20897$. In other words, these generators have the property of HELP.

© The Author(s), under exclusive license to Springer Nature Switzerland AG 2025
L. Deng et al., *Random Number Generators for Computer Simulation and Cyber Security*, Synthesis Lectures on Mathematics & Statistics, https://doi.org/10.1007/978-3-031-76722-7_14

Another type of PRNGs having the HELP property is MT19937 (proposed by Matsumoto and Nishimura 1998) and its related family of generators. MT19937 has the property of equi-distribution over dimensions up to 623, and perhaps is the most popular 32-bit PRNG presently. The underlying generating recurrence equation for MT19937 is an LFSR of order 19937 over the binary finite field.

Neither MRGs nor LFSRs (including MT19937) are secure generators, because one can compute their next value from a sequence of observed past values by solving a system of linear recurrence equations. See, for example, Stinson (2006, p. 37) for cryptanalysis of the LFSR stream cipher and Lidl and Niederreiter (1994, p. 235) for that of the MRG.

For pseudo-random bit generators (PRBGs), also known as deterministic random bit generators (DRBGs), Barker and Kelsey (2012) published NIST's recommendation and specification on approved DRBG algorithms, including those based on hash functions, block cipher algorithms (e.g., AES and Triple DES), and a number-theory problem (expressed in elliptic-curve technology). However, as pointed out in Stallings (2010, p. 234), these algorithms tend to be too slow for many applications that demand very high throughput. Another popular bit generator is the shrinking generator proposed by Coppersmith et al. (1994). Although the shrinking generator was shown to be vulnerable to the correlation attack (Golic 2001), it is an interesting example of using one LFSR acting as a (random) selector to choose or skip the output bit produced by another LFSR. Many other generators consider the strategy of applying a certain kind of non-linear filtering processes on multiple LFSRs; see, for example, Menezes et al. (1996). Some simple and elegant ciphers based on random table-shuffling have been proposed for computer security applications. RC4 is the best-known one tuned towards 8-bit architectures, while other ciphers such as HC-128/HC-256 are tuned towards 32-bit computer architectures. Note that HC-128, a variant of the original HC-256 proposed in Wu (2004), is one of the four algorithms in the eSTREAM software portfolio, see http://www.ecrypt.eu.org/stream/ and Robshaw and Billet (2008). Both RC4 and HC-128 are considered very efficient and practically secure.

In summary, up to now, two very different types of PRNGs have been developed in the literature and used in practice for applications in computer simulation and computer security, respectively, because of their totally different objectives and concerns. The major aim of this paper is to propose a general method to create a class of generators that are suitable for both applications. The general strategy of the method is to apply a non-linear mixing process to multiple PRNGs that have already enjoyed the HELP property. This new class of generators will be referred to as the Secure And Fast Encryption (SAFE) generators hereafter.

14.1   SAFE Generators

The proposed general mutual-shuffling method described earlier imposes no restrictions on the type of baseline generators used. Nonetheless, it would be of great advantages in security to have at least one of the two baseline generators be MT19937 or a large-order maximum-period MRG. To simplify the argument for the security analysis, we consider a SAFE generator constructed using two classical generators, MT19937 (for X) and MRG(k; m) (for Y), to serve as the baseline generators. Both generators have an extremely long period.

14.1.1 Key and IV Initialization Algorithm

We briefly address the issue of initialization for the internal states of X and Y. In theory, we can fill in the internal states and auxiliary tables with any secret keys and/or initialization vector (IV) values. Depending on the baseline generators used, the number of values that need to be initialized could be very large. For such circumstances, one can use a cryptographic primitive (e.g., DES, AES, or SHA) to create (e.g., for ANSI X9.17 generator) necessary initial values for the internal states and auxiliary tables based on one secret master key and IV values chosen by the user. Or, for simplicity, we can initialize each of the two shuffle tables separately by generating 2^C values from the associated baseline generators after the seed initialization. To add more complexity, we can also initialize the two shuffle tables with additional secret keys or by making use of any popular cryptographic primitives. Alternatively, we can use a similar procedure as in HC-128/HC-256 to first fill the initial states of the two baseline generators with the user key and given initialization vector, then fill the two auxiliary tables with the variates produced by the two baseline generators.

Algorithm 14.1: Generating Algorithm

---

1 At each iteration of generation, say, the i-th iteration, produce the output U_i as follows:

1. [**Internal states.**] Compute the current values of the internal states, X_i and Y_i, using the corresponding generating equations for the two baseline generators, respectively.
2. [**Intermediate states.**] Compute the values of the intermediate states, V_i and W_i, using the mutual-shuffling method with table-extraction and table-update as follows:

    a. Compute $V_i = T_x[r(Y_i)]$, and then $T_x[r(Y_i)] \leftarrow X_i$.
    b. Compute $W_i = T_y[r(X_i)]$, and then $T_y[r(X_i)] \leftarrow Y_i$.

3. [**Output.**] Output the i-th variate of U-sequence as

$$U_i \equiv G(X_i, Y_i; V_i, W_i), \tag{14.1}$$

where $G(X, Y; V, W)$ is one of the recommended choices as in Eq. 13.5 with bit rotation functions $f_j(x) = (x \gg r_j) \oplus (x \ll (32 - r_j))$ as in Eq. 13.1) using different rotation sizes r_j for $j = 1, 2, 3, 4$ on the values of internal and intermediate states.

14.2   Theoretical Property About SAFE

14.2.1 Property of Combination Generators

Consider two random-variate sequences, $\langle X_i \rangle$ and $\langle Y_i \rangle$, generated by two generators X and Y over $\mathbb{Z}_m$, respectively. Let $Z = (X + Y) \bmod m$, a combination generator of X and Y, and $\langle Z_i \rangle$ be the associated sequence. Let P_x, P_y, and P_z be the period lengths of generators X, Y, and Z, respectively. Denote the least common multiple and the greatest common divisor of P_x and P_y by $lcm(P_x, P_y)$ and $gcd(P_x, P_y)$, respectively.

Theoretical lower and upper bounds of P_z were provided in Li et al. (2012) as stated below in Theorem 14.1.

Theorem 14.1
$$\frac{lcm(P_x, P_y)}{gcd(P_x, P_y)} \le P_z \le lcm(P_x, P_y). \tag{14.2}$$

The combination generator Z is likely (but not always) to have a period length that can reach the upper bound $lcm(P_x, P_y)$. However, certain combinations may reduce its period length. For example, let $\langle C_i \rangle$ be the sequence generated by a generator of a period length P_c much shorter than P_x and P_y and let $Y_i = m - X_i + C_i$; then the combination generator becomes $Z_i = (X_i + Y_i) \bmod m$, which equals C_i, and thus has a period of length P_c. In particular, if we choose $Y_i = m - X_i + c$ for a constant $c \in \mathbb{Z}_m$, then $Z_i = c$, yielding a period of length 1. The latter example also shows that the lower bound given in Theorem 14.1 is a tight (i.e., reachable) lower bound. Generally speaking, the upper bound of $lcm(P_x, P_y)$ can be reached if the generating methods for $\langle X_i \rangle$ and $\langle Y_i \rangle$ are "unrelated". From statistical theory, X_i and Y_i being two statistically independent random variables would rule out the "bad" cases of $Y_i = m - X_i + C_i$. In fact, the upper bound is reached when $gcd(P_x, P_y) = 1$, for which $P_z = P_x \times P_y$ by Theorem 14.1. This is a well-known result given in Knuth (1998).

From the predictability aspect, if we are given only the value of Z_i, i.e., $(X_i + Y_i) \bmod m$, one cannot easily separate the values of X_i and Y_i. We formalize this fact in the following theorem.

Theorem 14.2 *Let X and Y be two independent random variables representing two random number generators. If X and Y are uniformly distributed over $\mathbb{Z}_m$ and let $Z = X + Y \bmod m$, then*

1. *Z is uniformly distributed over $\mathbb{Z}_m$. (This result still holds even when only X or Y is uniformly distributed over $\mathbb{Z}_m$.)*
2. *For any $z \in \mathbb{Z}_m$, the conditional distributions of $X|Z = z$ and $Y|Z = z$ are uniformly distributed over $\mathbb{Z}_m$.*

The above two theorems provide some security justifications for the combination method. If we are given only the value of Z, according to Theorem 14.2, the value of X or Y is equally likely to be any value in $\mathbb{Z}_m$. That is, given any $z \in \mathbb{Z}_m$, $P(X = x|Z = z) = 1/m$ for all $x \in \mathbb{Z}_m$ and $P(Y = y|Z = z) = 1/m$ for all $y \in \mathbb{Z}_m$. At each iteration of the generating process, the combination method produces one final output variate (Z_i) from the two internal variates (X_i and Y_i). Then, by carefully choosing generators X and Y, the combination generator $Z = (X + Y) \mod m$ would have a longer period (by Theorem 14.1) and better distributional property and unpredictability (by Theorem 14.2) than X and Y.

Theorem 14.2 provides a justification under the ideal (but unrealistic) assumption that both generators have the exact uniform distributions over $\mathbb{Z}_m$. As argued in Marsaglia (1985) and Deng and George (1990), the distributional property of the combination generator can be improved even when the two baseline generators are not uniformly distributed.

Furthermore, from Theorem 14.2, given only the value of Z_i, it is hard to recover X_i or Y_i. However, we should mention that adding two linear generators yields another linear generator; see, for example, L'Ecuyer (1996). Consequently, if we are given enough consecutive variates of the combination generator, it is easy to break the system by solving a set of linear equations. To avoid that, in this paper, we propose including at least one non-linear generator to form a combination generator; for example, $Z = (f_1(X) + f_2(Y)) \mod m$, where $f_1(\cdot)$ and $f_2(\cdot)$ are some efficient non-linear transformations discussed earlier.

To improve the generating efficiency of $Z = (X + Y) \mod m$, one can choose the modulus $m = 2^w$, where w is a positive integer. Then, $Z = (X + Y) \mod 2^w$ can be implemented as $(X + Y)\&(2^w - 1)$, where "&" is the logical "and" operation and $2^w - 1$ in its binary representation has w 1's in the w least significant bits and 0's in the rest of bits. The size of w in $m = 2^w$ is commonly chosen to be the CPU word size. In this paper, we fix $w = 32$, yet it can be easily increased to $w = 64$ or $w = 128$ for advanced CPUs, or decreased to $w = 16$ or $w = 8$ for simple CPUs.

14.2.2 Period of SAFE Generators

Having a large period length is one of the important factors in the cryptographic suitability of a secure PRNG, because a large number of states can eliminate the threat of the time-memory-data tradeoff attack on stream ciphers. We now investigate the period length of the proposed SAFE generators.

To simplify the analysis of SAFE generators, we consider the case of using two classical generators, MT19937 (for X) and $MRG(k; 2^{31} - 1)$ (for Y) with various values of $k \leq 20897$, to serve as the baseline generators. In particular, several DX-k generators are listed in Deng and Xu (2003) ($k = 102$ and 120), Deng (2005) ($k = 47, 643$, and 1597), and Deng et al. (2012b) ($k = 7499$ and 20897) with multipliers of the form $B = \pm 2^{r_1} \pm 2^{r_2}$. For these cases, $P_x = 2^{19937} - 1$ is relative prime to $P_y = m^k - 1$, where $m = 2^{31} - 1$. Therefore, according to Theorem 14.1, the exact period of the combination generator $Z =$

$(X + Y)$ mod m is $P_x \times P_y$. As to small-order MRGs, for $m = 2^{31} - 1$, we have found additional k's (less than 100) with known complete factorization of $m^k - 1$, including all $k \leq 24$ and $k = 26, 27, 30, 34, 39, 40, 42, 48, 51, 52, 60$. With complete factorization, it is then straightforward to find efficient maximum-period MRGs of small order k (available upon request).

The period length of the sequence $\langle U_i \rangle$ as defined in Eq. 14.1 depends on the choice of components in the mixing function $G(x, y; v, w)$ but not on the choice of non-linear bit-wise rotation transformations $f_j(\cdot)$, $j = 1, 2, 3, 4$, as long as they are one-to-one functions. No general formula is available for the period length of $\langle U_i \rangle$ and it would be tedious to discuss all 15 possible cases separately. For the recommended three-term mixing functions, we discuss the case of $G(x, y; v, w) = f_1(x) + f_2(y) + f_4(w)$ as an example. Recall that the combination generator with $(f_1(X_i) + f_2(Y_i))$ mod m has the period length of $P_x \times P_y$. For the generator with $(f_1(X_i) + f_2(Y_i) + f_4(W_i))$ mod m, even when $W_i = c$ (an unlikely event), its period length remains as $P_x \times P_y$. However, its period could be reduced when $f_4(W_i)$ is either $m - f_1(X_i) + c$, $m - f_2(Y_i) + c$, or $m - (f_1(X_i) + f_2(Y_i)) + c$, for all $i \geq 0$. Since W_i is a "shuffled" variate of *past* Y's taken from buffered table $T_y[\cdot]$, these conditions are unlikely to hold for all $i \geq 0$. Therefore, its period length is $P_x \times P_y$ in "most situations of practical interest".

Our recommended SAFE generators, with $G(x, y; v, w)$ involving three terms from $(x, y; v, w)$, should have the period length reaching the upper bound $P_x \times P_y$ using a similar argument as above. Generally speaking, the SAFE generator will start to repeat itself when both of the internal states of X and Y and the content of their shuffle tables, $T_x[\cdot]$ and $T_y[\cdot]$, are starting to repeat at the same time. One conservative lower bound for its period length is $\min(P_x, P_y)$, which is still very large for the two baseline generators recommended. For example, with MT19937 and a large-order DX-k (say, k=1597 or beyond) as baseline generators, this conservative lower bound is 10^{6001}, which is much longer than the "estimated period" of HC-128/HC-256. In practice, the potential problem of an extremely short period can be easily detected by applying extensive empirical tests on the generator.

From these discussions, we can infer that the period length of each proposed SAFE generator is long enough to guard against the time-memory-data tradeoff attack.

Since the number of the potential seeds (not all zeros) is huge for X as well as for Y, attacks utilizing an exhaustive search by adversaries are computationally infeasible. In the following, we provide some additional justifications for the security of the proposed SAFE generators.

14.2.3 Equi-Distribution Property and NNP Property

As mentioned earlier, a maximum-period MRG of order k has a nice equi-distribution property over dimensions up to k. We will use this property to support our claim that it is hard to predict the next variate if we have fewer than k consecutive past values.

Next Bit Predictors and Next Number Predictors

To study the security property, it is common to evaluate PRBGs via a useful concept called the *next bit predictor* (NBP) that predicts the n-th bit based on the previous $n - 1$ bits (see, e.g., Menezes et al. 1996, Stinson 2006, Yao 1982). A good bit generator should not have an NBP that can correctly predict the n-th bit of the generated bit-string with an additional accuracy from a blind guess (i.e., with a probability of 1/2) by more than ϵ except for very small $\epsilon > 0$. For a truly random bit generator, the size of ϵ for any NBP can be arbitrarily small. In practice, we only expect ϵ to be small enough so that there is no significant advantage to predict the next value from past values for any NBP. According to Yao (1982), NBP is a *universal test* for the security of PRBGs.

Extending from PRBGs to PRNGs, we consider, parallel to NBP, a concept called the *next number predictor* (NNP); and a good PRNG with modulus m should not have an NNP that can correctly predict the nth variate from the past $n - 1$ values with a probability larger than $1/m$ by any $\epsilon > 0$ except for very small ϵ.

NNP Property for Large-Order MRGs

For a maximum-period MRG of order k and modulus m, let $T_{m,k}$ be the sequence generated by the MRG. If $n \geq k + 1$ and we know all of its parameters, we can of course predict precisely its next value based on the previous $n - 1$ values. Conversely, when $n \leq k$, there is no way to predict its next value correctly based on the previous $n - 1$ values for a maximum-period MRG of order k. This can be justified by the following theorem.

Theorem 14.3 *For $n \leq k$, let $P_n(x|\mathbf{x}_{n-1})$ be the probability of observing $x \in \mathbb{Z}_m$ randomly chosen from the sequence $T_{m,k}$ given the previous $n - 1$ values, $\mathbf{x}_{n-1}$. If $\mathbf{x}_{n-1} \neq \mathbf{0}$, then $P_n(x|\mathbf{x}_{n-1}) = 1/m$. If $\mathbf{x}_{n-1} = \mathbf{0}$, then $P_n(x|\mathbf{x}_{n-1}) = 1/m + O(\frac{1}{m^{k-n+1}})$.*

Thus, under the case of $n \leq k$, no NNP exists with $\epsilon > E_0$. Clearly, E_0, as a function of $k - n$, decreases exponentially to zero in $k - n$, which means that it gets much harder to predict the nth number when the number of available past values gets smaller when compared to the order k. Consequently, when $n \leq k$, the chance of correctly predicting the next value (X_{j+n}) with the observed values of $(X_{j+1}, X_{j+2}, \ldots, X_{j+n-1})$ is hardly increased from $1/m$ (i.e., blind guessing) since the modulus m is a very large number, even under the assumption that all the parameters used in the MRG except for the seeds are made public to attackers.

For the case of $n > k$, it is true that Theorem 14.3 provides no guarantees on passing the "next-number test" for an MRG. Nevertheless, the MRG is just a baseline generator in the SAFE design and we believe the variates produced by the baseline generators are (computationally) well-protected through table shuffling, non-linear transformations, and combination/mixing.

From Theorem 14.3, it is clear that, the maximum-period MRG should have a large order k and/or a large modulus m to pass the "next-number test" (parallel to the next-bit test for PRBGs). For a small order k and a small modulus m, it may yield a small period length $m^k - 1$. In such a circumstance, if we further assume that one can observe past observations produced from the MRG, the equi-distribution property could be a disadvantage because one can design an attack by counting the frequency distribution of the observed variates. Since past variates of the baseline generators are hidden in the SAFE design and large values of order k and modulus m are used, it is cryptographically infeasible to take advantage of the equi-distribution property for next-number prediction; an adversary would need an astronomical amount of time even to read enough of the PRNG outputs to carry out the proposed attack.

14.3   Security Analysis

As required for any secure generators, the security of the proposed SAFE generators does not rely on the secrecy of the generating equations of the baseline generators; therefore, the SAFE generators are safe from the traditional attacks (e.g., Hastad and Shamir 1985; Boyar 1989) on finding the parameters of the targeted generator. In fact, the security strength relies on the difficulty in recovering the hidden values of the *current* "internal/intermediate states". The recovery is difficult because (i) the observable output is a non-linear transformation of the hidden "internal/intermediate states"; (ii) the parameter space for the *initial* internal states can be huge (up to 20897 words) when large-order MRGs are used for the baseline generators; (iii) to attackers, the hidden states are "moving targets" due to continual updating; and (iv) if large-order maximum-period MRGs are used for the baseline generators, the property of equi-distribution can ensure the equal likelihood for each of the huge possible "internal states"; hence, naive brute-force attacks cannot succeed with the current computing technology.

Our SAFE generators use various techniques to improve both distributional and security quality over the two classical baseline generators that have the nice property of HELP. The mutual-shuffling method and additional nonlinear transformations with the mixing function can break up the linearity structure of the baseline generators. Generally, the combination method (via the mixing function here) can further improve the quality of the generators. Combining two or more pseudo-random sequences into a "more uniform" random sequence was recommended by many researchers for the classical generators such as LCGs or small-order MRGs; see Gentle (2003), Marsaglia (1985), Knuth (1998, p. 35), L'Ecuyer (1996, 1999).

14.3.1 Against Certain Security Attacks

Recall that, for an MRG($k; m$) to achieve the maximum period of $m^k - 1$, any nonzero k-tuple $(s_1, s_2, \ldots, s_k)$, $s_i \in \mathbb{Z}_m$, will appear exactly once over its entire period. Therefore, it is infeasible to use brute-force methods to guess the internal-state values when m and k are large.

Given that the total number of possible design combinations of the generators, multipliers, index-selecting functions, table sizes, and output functions can be huge, plus the extremely long period lengths of the baseline generators when m and k are large, brute force methods of attacks are infeasible, if these parameters are not made public. However, a secure generator has to be able to guard against attacks when all the details about the generator are disclosed to attackers except for the private key.

Correlation attacks and algebraic attacks are known to be effective on most designs that use linear PRNGs (see Meier and Staffelbach 1989, Johansson and Jönsson 2000). The SAFE generator uses a non-linear mixer to break the linearity of the baseline generators, which is an effective defense for correlation attacks. More specifically, the use of the auxiliary tables to produce output from (random) selection functions and the use of non-linear output functions in the SAFE design are features similar to that of the well-known RC4 and HC-128 stream ciphers. Wu (2004) argued that these features make HC-256 (and therefore its variant HC-128) resistant to various classical attacks such as correlation attacks, algebraic attacks, linear attacks, and differential attacks. Each of our proposed SAFE generators not only has two large secret tables updated constantly as HC-256 does, it has two additional large secret internal states that are also updated at each iteration. Therefore, we can infer that the proposed SAFE generators are at least as secure as HC-256/HC-128; specifically, they are resistant to the aforementioned classical attacks.

An effective security against the time/memory/data tradeoff attack is to have a large number of states. Our SAFE generators can easily handle this by choosing a sufficiently large table size for the two shuffle tables. Also, as in Hong and Sarkar (2005), during the initialization, choosing a size for the initialization vector IV to be greater than or equal to the key size would provide a further barrier against the time/memory/data tradeoff attack.

14.3.2 Various Levels of Security/Defense

There are at least three layers of security/defense for our proposed SAFE generators:

1. [**Outer layer defense: hiding intermediate and internal states via coupling**] Recall that $U_i \equiv G(X_i, Y_i; V_i, W_i)$ is the output and $\{V_i\}$, $\{W_i\}$, $\{X_i\}$, and $\{Y_i\}$ are all hidden. Therefore, it appears difficult to recover these hidden values even from a large quantity of past values of the output state $\{U_i\}$.

2. [**Middle layer defense: hiding the ordering of internal states via shuffling**] Assume the worst-case scenario of a total breakdown on the outer layer that attackers have acquired all the values of the intermediate states, $\{V_i\}$ and $\{W_i\}$, which are scrambled values of the internal values, $\{X_i\}$ and $\{Y_i\}$, respectively. When the orders of the baseline generators and/or the shuffle table size (2^C) are large, it is computationally hard to restore the original ordering of the internal-state values, which are required for predicting future output values.

3. [**Inner layer defense: using large-order MRGs for their equi-distribution property**] Let us consider a scenario of a "partial" breakdown on the outer and middle layers that, for baseline generators X and Y, attackers only know fewer than their orders (i.e., 19937 and k) consecutive past internal-state values of X and Y, respectively. Using the high-dimensional equi-distribution property described in Sect. 14.2.1 and the next-number-predictor property to be described in the next subsection, the system can be proven still hard to predict under this scenario. When attackers know sufficiently many past values of the internal states in correct order and the current contents of the intermediate states (shuffle tables), the system would have a total breakdown. Although this situation is unlikely to happen, we may prevent such a total system breakdown by implementing a preventive "reseeding" scheme. For example, whenever a certain number of output values have been produced, change the values of the internal states and/or intermediate states (shuffle tables) following the seed construction process for reseeding as suggested by Barker and Kelsey (2012). The frequency of "reseeding" depends on one's belief on the security of the two outer layers described above. See, Barker and Kelsey (2012) for a discussion on this issue.

15.1 Previous (Failed) Attempts to Extend RC4

A number of attempts have been made to strengthen RC4, notably Spritz, RC4A, VMPC, and RC4+. Paul and Maitra (2011, Chap. 9) gave an extensive review and analysis of RC4 and its variants. In particular, they recommended RC4$^+$, proposed by Maitra and Paul (2008). According to Paul and Maitra (2011), KSA$^+$ (Key Schedule Algorithm) of RC4$^+$ can circumvent the weakness of standard RC4 KSA and PRGA$^+$ (RNG Algorithm) can better protect the information about the S-table by masking the output bytes. On the negative side, its running time is about 2.94 times (KSA$^+$) and 1.70 times (PRGA$^+$) of RC4 counterparts. In addition, RC4$^+$ is still a byte-oriented stream cipher.

We will only mention some other RC4 variants that are related to our proposed eRC. To expand 8-bit architecture of RC4 to 32-bit or 64-bit, Nawaz et al. (2005) and Gong et al. (2005) proposed RC(8, 32) and RC(8, 64), respectively. Their main idea is to fill the initial S-table with 32-bit (or 64-bit) numbers during their KSA with a very slight change in the PRGA. However, Wu (2005) showed that there is a simple distinguished attack on RC(8, 32) using only 100 key stream words. While the motivation of extending RC4 to a 32-bit stream cipher is useful, we believe that lack of continual injection of "random source" to the S-table is a major cause for the failure of their approach.

eRC: A General Scheme to Enhance/Extend RC4

RC4, its variants, and HC-128 share one common "weakness": they lack a "good" and "random" mechanism to drive the table-index selection in the algorithm. For RC4 and some of its variants, user keys are used to "scramble" the initial S-table, which may yield a "poor" permutation (randomization) of the initial S-table, especially when a weak KSA is used. Likewise, HC-128 requires a long iteration process to ensure that the initial tables are properly "randomized". The S-table of RC4 is always a "shuffle" of $0, 1, \ldots, 255(= 2^8 - 1)$. In contrast, the P/Q tables of HC-128 have less "justification" for their uniformity. Once the initial tables (S, P/Q) are fixed, the "randomness" of the output is produced through a sequence of complex transformations and combinations, without a continual "injection" of good external generators.

15.2   eRC: Adding a Good External RNG to RC4

As mentioned earlier, to enhance RC4, Deng et al. (2021) proposed adding an efficient RNG in the eRC design. Consider a general RNG that generates the sequence $\{X_i, i \geq 0\}$ by

$$X_i = f(X_{i-1}, \ldots, X_{i-k}) \bmod m, \quad i \geq 0, \tag{15.1}$$

with a "suitably chosen" integer-valued function, f, of the most recent k integers in the past and k initial seeds, $X_{-k}, X_{-(k-1)}, \ldots, X_{-1}$. We shall consider a specific choice of the baseline generator later.

There are several advantages of adding a good RNG as a baseline generator to RC4: (i) better initialization of the S-table with variates produced by the baseline generator, (ii) continuous updates of the entries in the S-table via an external RNG, (iii) better random index selection of the S-table, (iv) enhanced security property due to large and constantly updated internal states from the baseline generator, (v) good statistical properties of the generated variates of eRC due to the initialization and constant updates of the S-table by the baseline generator, and (vi) higher throughput for eRC compared to RC4 (32 bits versus 8 bits) at each iteration, implying, the relative efficiency "gain" of eRC over RC4 can be achieved by using a faster baseline generator.

Next, we propose a general framework that can be viewed as an *"enhancement"* for RC4. The key idea is to add an external generator that continuously feeds/updates the S-table to ensure better uniformity and randomness.

General framework to extend RC4

Like RC4, let S be a table of size 2^C; but unlike RC4, each entry of the S-table holds a value generated by the baseline generator, which can be of any size, say, 32-bit, 64-bit, or even larger.

Consider the general framework of eRC as shown in Fig. 15.1 (at the ith iteration). We briefly describe each of the blocks in the diagram:

1. To begin the generation step, we need to initialize the internal states of RNG-X (the baseline generator), the contents of the S-table, and the initial values of indices I and J. All of these need to be filled using the private user-keys and initialization values (IVs).

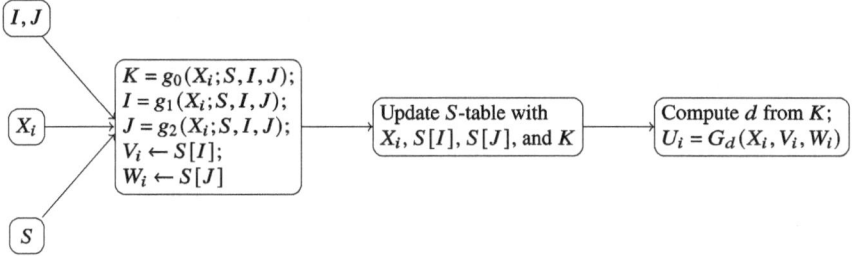

Fig. 15.1 Diagram of eRC method

At the beginning of the ith iteration, we have the current values of I, J, X_i, and the S-table.

2. Compute the new values of I, J, and K based on the current values of I, J, the S-table, and X_i. The purpose of I, J, and K in eRC are similar to that of RC4:

   a. $K = g_0(X_i; S, I, J)$ will produce one random number using a combination of the current variate (X_i) and previous $S[I]$ and $S[J]$ at the beginning of the iteration. Note that K is computed from feedbacks (i.e., the $S[I]$ and $S[J]$ from the previous iteration).

   b. $I = g_1(X_i; S, I, J)$ and $J = g_2(X_i; S, I, J)$ are suitably chosen "C-bit generators" producing the selection indices used for choosing entries from the S-table.

3. Different bits (say, $a(K)$ and $b(K)$) of K will be used to (a) determine the location to store X_i in the table $S[\cdot]$ (using $a(K)$) and/or (b) randomly select one mixing function from a list of possible 2^D functions (using $b(K)$):

   a. Use $a(K)$ to update the S-table with some "random shuffling" of the S-table and "random assignment" of X_i into the S-table. In RC4, the S-table (of size 2^C) is "shuffled" by swapping $S[I]$ and $S[J]$. In our design, in addition to "shuffling", we need a scheme to randomly update $S[I]$ or $S[J]$.

   b. Use $b(K)$ to perform a random selection from a list of 2^D output transformations, say, $G_d(X_i, V_i, W_i), d = 0, 1, \ldots, 2^D - 1$. This new feature of random selection of the output function can greatly increase the difficulty of mounting attacks.

   c. It is easy to show that if random variable K is uniformly distributed over $0, 1, \ldots, 2^w - 1$, where w is a positive integer, then the two random binary variables $a(K)$ and $b(K)$ are uniformly distributed; and they are statistically independent of each other. Therefore, we can make "independent" decisions based on different parts of K.

For each of the stages listed above, many choices for these procedures can be considered and we will provide our recommendations later. The main requirement is that the chosen procedures would enhance the desired properties of "randomness", efficiency, and unpredictability. Obviously, it is important to choose a "good" RNG of 32 bits (or higher) with an extremely long period length, great efficiency, and nice statistical properties to be the baseline generator. There are many such RNGs proposed in the literature recently, mainly for computer simulation applications.

15.2.1 eRC-E

There are many choices that can be used for table index selection to produce I, J, and K for the general eRC design. With simplicity and efficiency in mind, we consider specific choices that have strong resemblance to the popular RC4 cipher. Specifically, we let

1. $K = (S[I] + S[J]) \gg A$, where A is a chosen shift size. Usually, we would choose $A \geq 1$ to avoid the situation where the least significant bit of $S[I] + S[J]$ is slightly biased, especially when the table size 2^C is small. Therefore, the selection of K is similar to but not quite the same as RC4.
2. $I = I + 1 \bmod 2^C$ and $J = J + S[I] \bmod 2^C$, same as RC4.

Given that we are using a similar index scheme as RC4, we shall call the generators based on the following eRC design simply eRC generators (Fig. 15.2). Where $a(K)$ and $b(K)$ are binary variates from different bits of K as discussed in the general eRC design.

Two proposed procedures to update S-table in eRC design.

As mentioned earlier, we need to consider proper procedures to "shuffle" the S-table in a manner similar to RC4 and update the contents of the S-table with X_i. The following are some simple procedures that can be considered:

1. First let $S[I] \leftarrow S[J]$ and then $S[J] \leftarrow X_i$. This is clearly very straightforward and efficient but it lacks "symmetry" between $S[I]$ and $S[J]$. Note that it may not be a good

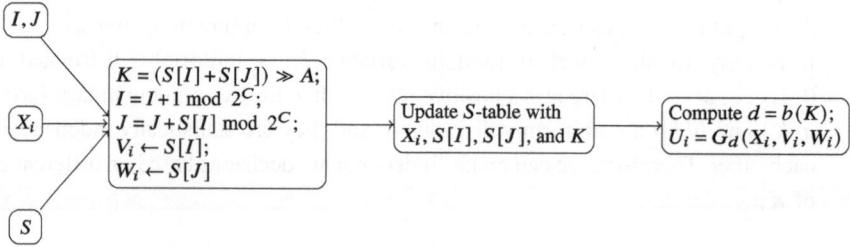

Fig. 15.2 Diagram of simple eRC method

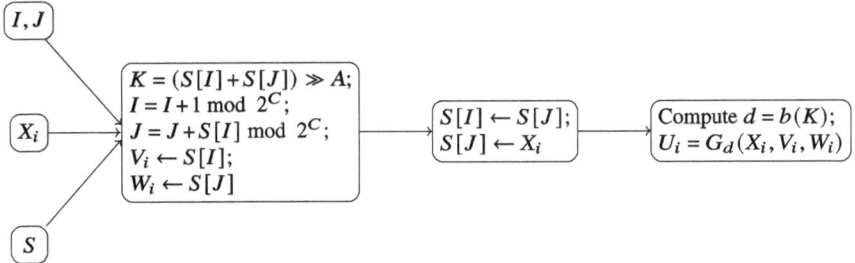

Fig. 15.3 Diagram of eRCe method

idea to use "$S[J] \leftarrow S[I]; S[I] \leftarrow X_i$" since $I = I + 1 \bmod 2^C$, with which the entries in the S-table may thus be updated sequentially.

2. We can also use a bit of K ($a(K)$) to decide the updating scheme for the S-table. For example, if $a(K) = 0$ then let $S[J] \leftarrow S[I]$ and then $S[I] \leftarrow X_i$; else let $S[I] \leftarrow S[J]$ and then $S[J] \leftarrow X_i$. This procedure is more "secure" (with random decision) and it is "symmetric" between $S[I]$ and $S[J]$. However, it is less efficient than the first proposed procedure.

3. To further increase the complexity of the updating scheme, we can use multiple (2 or more) bits of K to make a random selection from various (reasonable) updating schemes for the S-table. In this sense, the two procedures described above are its special cases. However, it can be less efficient than the two aforementioned procedures.

For the design with the first procedure to update the S-table, we denote it by $eRC(RNG - X; E, 2^C, 2^D)$, where "E" stands for efficiency. It can be described in Fig. 15.3.

For the design with the second procedure to update the S-table, we use **eRC(RNG − X; S, 2^C, 2^D)** to denote it, where "S" stands for symmetry (or security). It can be described in the diagram in Fig. 15.4.

15.2.2 eRC-**S**

As shown later, with a good baseline generator, both procedures produce RNGs that can pass stringent empirical tests.

Choice of output functions

We define the list of output transformations via combinations of several bit rotation functions as defined below:

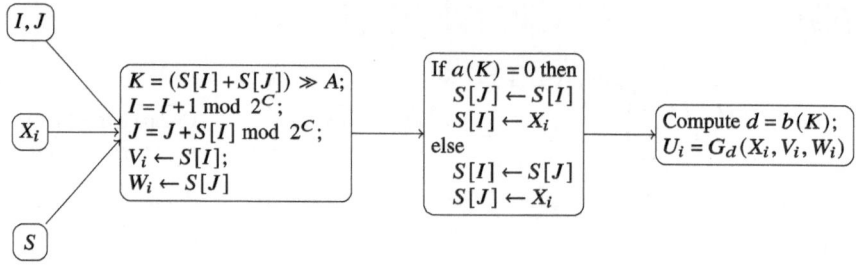

Fig. 15.4 Diagram of eRCs method

1. The bit rotation function for a 32-bit variate x is defined as

$$f_r(x) = (x \gg r) \oplus (x \ll (32 - r)),$$

   where r is the size of bit rotation. The bit rotation function is similar to the ones used in Wu (2004).

2. For the list of 2^D output transformation functions, we consider

$$G_d(x_i, v_i, w_i) = f_{s_{d1}}(x_i) \oplus f_{s_{d2}}(v_i) \oplus f_{s_{d3}}(w_i),$$

   with $0 < s_{dj} < 31$ for $j = 1, 2, 3$ chosen with different values for the bit rotation sizes, $d = 0, 1, \ldots, 2^D - 1$. For $D = 2$, we can choose from the four possible output functions below:

   a. $G_0(X_i, V_i, W_i) = f_{s_{01}}(X_i) \boxplus f_{s_{02}}(V_i) \boxplus f_{s_{03}}(W_i)$.
   b. $G_1(X_i, V_i, W_i) = f_{s_{11}}(Y_i) \boxplus f_{s_{12}}(V_i) \boxplus f_{s_{13}}(W_i)$.
   c. $G_2(X_i, V_i, W_i) = f_{s_{21}}(X_i) \boxplus f_{s_{22}}(Y_i) \boxplus f_{s_{23}}(V_i)$.
   d. $G_3(X_i, V_i, W_i) = f_{s_{31}}(X_i) \boxplus f_{s_{32}}(Y_i) \boxplus f_{s_{33}}(W_i)$.

   Here we assume that different shifts $1 \le s_{dj} \le 31$ ($d = 0, 1, \ldots, 2^D - 1$, $j = 1, 2, 3$) are used; therefore there are plenty of possible different configurations.

   The list size, 2^D, does not have to be large and we set $D \le 2$. When $D = 0$, with a single output function, it is the most efficient but may lack a degree of complexity.

15.3   Design Characteristics

The design principles as well as the characteristics of the design are described below.

1. A baseline generator is chosen from a huge pool of good generators developed recently for computer simulation applications. See Deng and Bowman (2017), Deng and Shiau (2015) for a review. Depending on the security requirements, we can choose a baseline generator that has a suitable size for the *initial* "internal states" (i.e., the seeds of the generator). The *current* "internal states" are continually updated by successive outputs from the baseline generator.

2. Like RC4, eRC uses a S-table of reasonable size 2^C (say, up to 512). Unlike RC4, the S-table of eRC holds the generated values from the baseline generator, which can produce a variate with a specific number of bits (say, 32-bit, 64-bit, or even larger). While RC4 can produce only C (usually $C = 8$) bits at each iteration, the output size of eRC does not depend on the size of C and it can generate variates of size 32 (or beyond). Therefore, if we choose an efficient RNG as the baseline generator, then the eRC generator is expected to be more efficient than the popular RC4. This is because RC4 requires 4 iterations to produce a 32-bit output and eRC needs only one iteration running essentially similar steps. This observation was verified later using actual empirical evaluations.

3. Both RC4 and the two eRC generators use similar formulas to produce the needed indices, I, J, and K. They are then used to select (using I and J) corresponding entries from the S-table or to produce output (using K).

4. For both RC4 and eRC, the output variate is produced using the current values of I, J, and the S-table. RC4 uses them to obtain another index $K = S[I] + S[J] \bmod N$ to select the output entry from the S-table. On the other hand, a general eRC mixes the two table entries indexed by I and J at each iteration. Any mixing function can be used, as long as it is hard to trace back to the individual hidden values while retaining good distributional properties.

5. Both RC4 and eRC continually update the contents of the S-table. For RC4, $S[I]$ and $S[J]$ in the S-table are swapped. For eRC, the current value generated by the baseline generator is used to replace the I-th entry in the associated S-table.

6. In the eRC design, random variate K is computed based on the hidden feedback (i.e., $S[I]$ and $S[J]$) from the previous iteration. The use of "feedback" as an 'input' for the next iteration has been used in Bays and Durham (1976) to enhance a "poor" linear RNG with a non-linear structure between output variates; and it can also greatly increase the period of a RNG with a large table size 2^C. However, there is no known formula for the exact relation between the period length and its table size.

7. The output variate is produced by an output function selected randomly from a list of possible non-linear transformation functions, $G_d(X_i, V_i, W_i)$, $d = 0, 1, \ldots, 2^D - 1$, using D-bits from the index selector $K = g_0(X_i; S, I, J)$. The choice of the list size 2^D is flexible, depending on the security and efficiency requirements. Large table size in general can improve the complexity of the proposed cipher and hence increase the

difficulty of algebraic attacks. When $D = 0$, it is reduced to a simpler design with a fixed non-linear transformation function.

While using a baseline generator in eRC can help improve/enhance RC4, using the S-table in the RC4-like design can also help improve the performance of a linear generator in terms of its unpredictability and/or its statistical/distributional properties. The linearity structure of the baseline generator can be broken by the use of the S-table in the eRC scheme. The idea of shuffling using a table buffer is not new in the computer simulation community. For improving the empirical performance of LCGs, MacLaren and Marsaglia (1965) proposed using an auxiliary LCG to shuffle the output generated by a baseline LCG. The S-table in the proposed eRC scheme, serving as a buffer holding the generated values of the baseline generator, also has the effect of breaking the linear relationship among successive generated.

A method is proposed to improve the security of the HC-128 stream cipher. HC-256 is one of the four software-based stream ciphers selected in eSTREAM Portfolio (http://www.ecrypt. eu.org/stream/), and HC-128 is a small variant of HC-256 with two secret tables (P/Q), each with 512 32-bit integers, 128-bit key (**K**) and 128-bit initialization vector (**IV**). HC-128 is considered to have a strong security property and fairly efficient key generation. The main drawback is that it needs a long time for the initialization with the **K** and **IV** setup algorithms to expand into P/Q tables. In addition, there is no known theoretical justification for its distributional property and no known result about its period length. In this chapter, eHC(RNG_X, RNG_Y), an enhanced HC-128 with two external generators is presented: RNG_X, RNG_Y are added to the original HC-128. The **K** and **IV** are used to initialize the seeds needed for the generators. It then initializes the P/Q tables by good external generators. To enjoy the security property of HC-128, eHC makes minimal changes during each step of the keystream generation stage with a simple mixing of two external generators.

16.1 HC-128 and HC-256 Stream Ciphers

The keystream generation is done by the following algorithm. A few comments about the keystream generation procedure:

1. All entries of P/Q tables are updated sequentially with some ARX operations.
2. The output $s_i = h_1(P[j \boxminus 12]) \oplus P[j]$ a simple coupling (XOR) with two entries in P where $x \boxminus c$ denote $x - c$ mod 512,
3. $s_i = h_2(Q[j \boxminus 12]) \oplus Q[j]$ a simple coupling (XOR) with two entries in Q.

© The Author(s), under exclusive license to Springer Nature Switzerland AG 2025   215
L. Deng et al., *Random Number Generators for Computer Simulation and Cyber Security*, Synthesis Lectures on Mathematics & Statistics,
https://doi.org/10.1007/978-3-031-76722-7_16

4. The exact period length of HC-128 is highly dependent on the Key/IV values and it is hard to theoretically analyze and compute its period length.

Algorithm 16.1: HC-128 Keystream generation algorithm

---

1 $i = 0$

2 **repeat**

3     $j = i \bmod 512$;

4   **if** *(i mod 1024) < 512* **then**

5        $P[j] = P[j] + g_1(P[j \boxminus 3]; P[j \boxminus 10]; P[j \boxminus 511])$

6        $s_i = h_1(P[j \boxminus 12]) \oplus P[j]$

7   **end**

8   **else**

9        $Q[j] = Q[j] + g_2(Q[j \boxminus 3]; Q[j \boxminus 10]; Q[j \boxminus 511])$

10        $s_i = h_2(Q[j \boxminus 12]) \oplus Q[j]$

11        $i = i + 1$

12   **end**

13 **until** *until enough keystream bits are generated*;

16.2   Generation Speed of HC-128

The speed of generation for stream ciphers is highly dependent on the platform, CPUs and compilers used. Generally speaking, comparison of the speed of generation may not be able to provide a definite conclusion. Bernstein, D.J. http://cr.yp.to/streamciphers/timings.html using three different CPUs: (a) Intel Core 2 Duo, (b) AMD-Athlon X2, and (c) Motorola PowerPC G4 with eSTREAM timing tool compared the number of cycles consumed per byte on several stream ciphers. We report only CryptMT3, HC256, SOSEMANUK, Salsa20 and AES (counter-mode). CryptMT3 was among the eSTREAM project competitions but not selected as final four stream ciphers, though it is the fastest in generation on Intel Core 2 Duo CPU, reflecting the efficiency of SIMD operations in this newer CPU. However, CryptMT3 is slower on Motorola PowerPC. This is because AltiVec (SIMD of PowerPC) lacks 32-bit integer multiplication (so we used non-SIMD multiplication instead). Matsumoto et al. (2007, 2008) proposed CryptMT version 3 (CryptMT3) to implement the original design on some modern CPUs (e.g., Intel Core 2 Duo) with SIMD (single-instruction-multiple-data) operations on 128-bit registers. They reported that CryptMT3 is about 1.8 times faster than the original CryptMT and can be considered as one of the fastest stream ciphers.

A subset of data, copied from Bernstein's page, is listed in Table 16.1.

From Table 16.1, we can see that HC-256 takes a much longer time for Key/IV setup. Given HC-128 is a smaller variant of HC-256, its initialization time is expected to be still the highest. The distribution of initial P/Q tables (from step 2 to step 3 in Algorithm 16.1 above) may depend heavily on the values of the key and IVs selected. Therefore, the table

Table 16.1 Summary from eSTREAM benchmark by Bernstein's page

Primitive	Core 2 Duo		AMD Athlon64 X2		PowerPC G4	
	Stream	K/IV setup	Stream	K/IV setup	Stream	K/IV setup
CryptMT3	2.95	575.83	4.73	612.64	9.23	823.51
HC-256	3.42	83866.64	4.26	88831.31	6.17	71479.71
SOSEMANUK	3.67	1473.50	4.41	1657.82	6.17	2387.50
Salsa20	7.12	34.33	7.64	112.31	4.24	111.93
AES-CTR	19.08	644.34	20.42	955.65	34.81	339.92

initialization requires 1024 steps (4–8), of non-linear ARX operations. There are other key disadvantages:

1. **Relatively large table sizes (512 words each) required**. The table sizes of P/Q in HC-128 are not flexible without much theoretical justification. On the other hand, smaller table sizes may weaken its security.
2. **No known theoretical justification about its distribution or period length**. The distribution property and/or period length of HC-128 are unknown and they may depend heavily on the keys/IVs chosen.

From Table 16.1, we can clearly see that HC-256 stream generation speed is highly competitive. Given HC-128 is a smaller variant, it is expected to be faster. After the long successful table initialization, the key generation step is reasonably efficient. Therefore, HC-128 could be useful for generating a long sequence of stream cipher. The other key advantage of HC-128 is its strong security property. It is generally recognized that HC-128 is a secure cipher and as mentioned before, its larger variant, HC-256, was chosen as one of the four finalists in eSTREAM project.

16.3   SAFE Generator and the Improvement to HC-128

We present a novel and simple enhancement to HC-128 in a later section. The idea is motivated by a proposed design called SAFE (Secure And Safe Encryption) in Sect. 14.1. The SAFE design included two good external generators. The authors proposed to initialize the initial seeds of the external generator from Keys/IVs, and then use them to fill two (much smaller) shuffle tables like P/Q table for HC-128/HC-256. Therefore, we can be sure that P/Q tables with desirable distributional properties can be made quickly. The key generation procedure of SAFE is different from HC-128 with the idea of mutual random shuffling, random update, and ARX operations. According to the empirical study, SAFE is as efficient as the well-known fast-stream cipher CryptMT3. Moreover, SAFE generators do not require the use of 128-bit registers with SIMD operations.

In order to describe the formal proposal, we need to give a brief review on the recent developments in classical pseudo-random number generation in the next section.

16.4   eHC: Enhanced HC-128

An enhanced HC-128—called eHC(RNG_X, RNG_Y)—is proposed, by introducing two good external generators that are used to fill in quickly the HC-128's P/Q tables in the table initialization stage, without the need for a long initialization. During the key generation stage, RNG_X and RNG_Y are used to combine and update the P/Q tables sequentially with minimal changes to the original HC-128 design. There are several main advantages of eHC:

1. **K** and **IV** are used to initialize the seeds of two external generators,
2. it can greatly shorten the time for initializing the P/Q tables,
3. it offers a much stronger justification for the uniformity property with any input **K** and **IV**, and
4. because of the (nearly) identical steps as HC-128, eHC should be as secure as HC-128, and the same security analysis can be applied directly to eHC.

The idea of enhancing HC-128 is to add two good external generators RNG_X and RNG_Y. Following similar notation as SAFE(RNG_X, RNG_Y), we call the new design eHC(RNG_X, RNG_Y). A simple and effective procedure to enhance HC-128, particularly its efficiency, without compromising its security, is as follows,

16.4.1 Initialization of P/Q Tables for eHC

Let $K = K_0 \| K_1 \| K_2 \| K_3$ be the 128-bit keys and $IV = IV_0 \| IV_1 \| IV_2 \| IV_3$ the 128-bit IVs.

Following Deng et al. (2018) for SAFE design, we can use a similar approach as an initialization step to fill in the internal state vectors for RNG_X and RNG_Y after some small burn-in cycles to mix secret keys and IV values. For security consideration, we can also apply a popular secure scheme (e.g. ChaCha or Salsa20 as in Bernstein 2008b) to mix the secret keys and IVs and then use its outputs to fill in initial state vectors of the baseline generators. After the internal state vectors are initialized, we can sequentially fill in the two auxiliary tables, P/Q, with variates produced from the RNG_X and RNG_Y where part of the secret keys and IV values were used in the initialization step.

There are several key features of the new initialization:

1. Key/IV are used to set the initial seeds of two external generators, not P/Q tables directly.
2. P/Q tables are filled directly from two external generators which should be much more efficient with a much nicer distributional property.

3. If needed, we can set a small number of runs as burn-in. The purpose is to further dilute the information about Key/IV.
4. Since the P/Q tables are updated by two good PRNGs, the number of iterations R in the burn-in period does not need to be large (say, 12 or less) and it can be controlled by the key and IV.
5. The period length of initialization is the least common multiple of the two period lengths of the generators used. There is great flexibility to choose the orders of the two generators. If needed, the size of Key and IV can be greatly increased for stronger security.

Algorithm 16.2: eHC-128 initialization algorithm

---

1. Input: $\mathbf{K} = K_0 \| K_1 \| K_2 \| K_3$ and $\mathbf{IV} = IV_0 \| IV_1 \| IV_2 \| IV_3$ and two wo external generators, RNG_X and RNG_Y of orders k and h
2. Applying an expansion scheme, expand $\mathbf{K}$ and $\mathbf{IV}$ into number of initial seeds (with a total of $k + h$) needed for RNG_X and RNG_Y.
3. Select the number of iterations R in the burn-in period does not need to be large (say, 12 or less) and it can be controlled by $\mathbf{K}$ and $\mathbf{IV}$.
4. Update the tables P and Q with the two external RNGs:
   for $i = 0$ *to 511* **do**
   $\quad P[i] = \text{RNG}_X()$
   $\quad Q[i] = \text{RNG}_Y()$
   end

There are several key points in Algorithm 16.2 for eHC initialization procedure:

1. There are several choices for mixing/expansion algorithms for **K/IV** that can be selected in steps 1–2 to expand the four 32-bit K and four 32-bit IV into an array of initial seeds for the two generators which required k and h initial works, respectively. One can choose the desirable mixing algorithm with a given security requirement.
2. Since any starting (non-zero) seeds will yield maximum period by the maximum period RNGs. Hence, Key/IV will not have any effect on the randomness or period length of the P/Q tables.
3. The order k and h can be highly flexible with extremely long period lengths for generating the P/Q tables. With enough randomness and uniformity of P/Q, there is no need to require additional ARX operations to mix (sequentially) the 1024 entries of P/Q tables.
4. Without using two external generators, the initialization of P/Q tables of HC-128 from Key/IV is a long and tedious process to ensure a reasonable uniform/random distribution property for P/Q tables. In our opinion, there is no theoretical justification for those ARX operations will achieve this property.
5. For eHC, **K/IV** were used to initialize the initial seeds of two baseline generators, RNG_X and RNG_Y. The P/Q tables can then be filled with the sequential outputs of

two generators which are known to produce random numbers with good distributional property. Therefore, eHC does not require a long initialization process like HC-128.

16.4.2 eHC Key Stream Generation

The key stream generation for eHC is done by the following algorithm with minimal additions of two generators to the HC-128.

Algorithm 16.3: eHC-128 Key stream generation algorithm

---

1 $i = 0$
2 **repeat**
3      $j = i \bmod 512$;
4   **if** *(i mod 1024) < 512* **then**
5          $P[j] = P[j] + g_1(P[j \boxminus 3]; P[j \boxminus 10]; P[j \boxminus 511]) + \mathrm{RNG}_X()$
6          $s_i = h_1(P[j \boxminus 12]) \oplus P[j]$
7   **end**
8   **else**
9          $Q[j] = Q[j] + g_2(Q[j \boxminus 3]; Q[j \boxminus 10]; Q[j \boxminus 511]) + \mathrm{RNG}_Y()$
10         $s_i = h_2(Q[j \boxminus 12]) \oplus Q[j]$
11         $i = i + 1$
12   **end**
13 **until** *until enough keystream bits are generated*;

Since HC-128 is highly regarded as a stream cipher with a known strong security property and fast generation speed, it would make sense to make a minimum change to the original design of HC-128.

1. Note that line 5 and line 9 in the above eHC algorithm are slight modifications from the original HC-128 algorithm with addition (or combination) of $\mathrm{RNG}_X()$ and $\mathrm{RNG}_Y()$, respectively. It is well known that combination can improve the distributional property.
2. eHC key stream generation is reduced to HC-128 if we choose $\mathrm{RNG}_X()$ and $\mathrm{RNG}_Y()$ to zero. Therefore, eHC should be as secure as HC-128, if not more.
3. After P/Q initialization, HC-128 will produce random numbers via a series of ARX operations without external random sources. In contrast, the P/Q tables in eHC are continuously updated with the external generators.

Remarks

We have proposed eHC, a simple but effective enhancement of the popular HC-128 and we summarize the main advantages of eHC over HC-128:

1. Comparing with HC-128, eHC should have better distribution properties with combinations of external generators which can improve the uniformity property of the P/Q tables.
2. Compared with HC-128, eHC has much a much faster initialization of the P/Q tables.
3. eHC has great flexibility to choose a good external generator to greatly increase the period length of eHC.
4. In addition to the HC-128 ARX operations, continuous updates of P/Q tables with good external generators are essential to ensure the randomness/uniformity of the outputs.
5. Maintaining the security property with minor changes to the original design. It is easy to see that if RNG_X and RNG_Y are producing zero values, then eHC is reduced to HC-128. Therefore, we can argue that eHC is as secure (if not more) as HC-128 which has been accepted as the four finalists in eSTREAM project.

We can briefly compare designs of eHC and SAFE: Both used the same initialization procedure with two good external PRNGs to fill two tables. The key generation procedure is quite different: eHC used the slight modification of HC-128 key generation steps whereas SAFE used mutual shuffling with random tables updates with different ARX operations. One big drawback of eHC is that it is not straightforward to reduce the P/Q table sizes which were the same as HC-128. Therefore, eHC is not suitable as a hardware stream cipher whereas it is much more flexible/scaleable for SAFE RNG_X, RNG_Y) to choose the two baseline generators.

17.1 Cryptanalysis

For low values of R, Salsa20 and ChaCha have been broken (Aumasson et al. 2008; Crowley 2005; Ishiguro et al. 2011; Maitra 2015; Shi et al. 2013). Crowley (2005) first proposed a 2^{165} operation attack on Salsa20/5 based on truncated differential cryptanalysis. Since then there have been several attacks of the same type on Salsa20/6, Salsa20/7, and Salsa20/8 as well as ChaCha6 and ChaCha7. See Choudhuri and Maitra (2016b) for a detailed list. For Salsa20/5 and Salsa20/6, Choudhuri and Maitra (2016b) substantially reduced the complexity of attacks to a practical level using multi-bit differentials. They also found 2^6, 2^{12} and 2^{16} operation attacks on ChaCha 4, 4.5, and 5 respectively. Their method was not as successful on rounds of 7 and 8 for Salsa20 and rounds of 6 and 7 for ChaCha, reducing complexity but not to the point of practicality.

The full round of Salsa20/12 was a finalist in eSTREAM and is considered to be secure. A proof of Mouha and Preneel (2013) demonstrated that differential cryptanalysis on Salsa20/15 would be more difficult than 128-bit key exhaustion. There is no security proof of how many rounds are required to ensure the safety of either Salsa20 or ChaCha. Most designers will choose a high number of rounds to avoid potential attacks. This choice ensures higher safety but adversely affects the speed of the cipher. Choudhuri and Maitra (2016a) proposed a hybrid model to evaluate Salsa and ChaCha and investigate the number of rounds required for security. Their results indicated that a total of 12 rounds are sufficient to achieve an appropriate level of security.

As two successive initial state matrices of sixteen words, the difference is often 1 at the position of "t_0" unless there is a carry into t_1. In our opinion, this unique feature could be one of the main methods of attacks with differential cryptanalysis as previously reported. Therefore, it makes sense to have a larger value of R so that it would be harder for the attacker at the expense of generating efficiency. It is amazing that such a minor difference

L. Deng et al., *Random Number Generators for Computer Simulation and Cyber Security*, Synthesis Lectures on Mathematics & Statistics, https://doi.org/10.1007/978-3-031-76722-7_17

can make the two output blocks "look" unrelated to the attacker after a sequence of R rounds of transformations/scramblings. However, fast diffusion does not automatically imply a good "distributional property" (uniformity, independence, theoretical, and empirical) of the generated variates.

17.2   Previous Extensions of Salsa20 and ChaCha

Both Salsa20 and ChaCha produce a 64-byte block using a 256-bit key, 64-bit nonce and 64-bit block counter. Compared to other eSTREAM finalists, the 64-bit nonce (v_0 and v_1) is somewhat smaller. For example, HC-256, a variant of an eSTREAM finalist HC-128, has 256-bit nonce and Sosemanuk, another eSTREAM finalist, has 128-bit nonce. Larger size of nonce does not imply improved security, it simply allows more messages to be sent/processed under the same key. However, large nonce size can minimize the possibility of collision which is a feature suitable for "random" generation of nonces.

There are two simple and popular solutions to expand the nonce size. A simple variation of ChaCha allows for a bigger nonce by using three words (96 bits) for the nonce and only one for the counter (Nir and Langley 2018). That is, we replace one of the counters, t_1, in the state matrix in Eq. 12.1 with the new (third) nonce v_2 by simply letting $t_1 = v_2$ in Eq. 12.1. However, this simple procedure reduces the maximum keystream blocks from 2^{64} down to 2^{32}.

Another variation XSalsa20 that allows for a 192-bit nonce, denoted as (v_0', v_1', v_2', v_3', v_4', v_5') without sacrificing the counter or other components was proposed in Bernstein (2011). In principle one can design XChaCha using the same nonce extension technique with ChaCha, but (Bernstein 2011) didn't formally publish an exact specification.

The idea is to run the initial Salsa20 with slightly different initial state matrix than the one in Eq. 12.1 as

$$\mathbf{Y}_S^{(0)} = \begin{pmatrix} c_0 & k_0 & k_1 & k_2 \\ k_3 & c_1 & v_0' & v_1' \\ v_2' & v_3' & c_2 & k_4 \\ k_5 & k_6 & k_7 & c_3 \end{pmatrix}, \quad \mathbf{Y}_C^{(0)} = \begin{pmatrix} c_0 & c_1 & c_2 & c_3 \\ k_0 & k_1 & k_2 & k_3 \\ k_4 & k_5 & k_6 & k_7 \\ v_2' & v_3' & v_0' & v_1' \end{pmatrix}, \qquad (17.1)$$

with the original counter (t_0, t_1) replaced by (v_2', v_3') (two new 32-bit nonce). Applying Salsa20/ChaCha transformation as

$$\mathbf{Y}_S^{(R)} = \mathbf{H}_S(\mathbf{Y}_S^{(0)}; R), \qquad (17.2)$$

and

$$\mathbf{Y}_C^{(R)} = \mathbf{H}_C(\mathbf{Y}_C^{(0)}; R). \qquad (17.3)$$

This procedure should be able to mix the keys and IVs well in the resulting matrix (vector) of $\mathbf{Y}_S^{(R)}$ (or $\mathbf{Y}_C^{(R)}$) which will be denoted as a vector of

$$\mathbf{K} = (z_0, z_1, \ldots, z_{15}).\tag{17.4}$$

The general idea is to replace the 256-bit keys $(k_0, k_1, \ldots, k_7)$ (eight 32-bit words) with prescribed eight words from that vector $\mathbf{K}$. Specifically, for XSalsa20, Bernstein (2011) chooses

$$(k_0', k_1', k_2', k_3', k_4', k_5', k_6', k_7') = (z_0, z_5, z_{10}, z_{15}, z_6, z_7, z_8, z_9)$$

for XChaCha, no specification by the author of ChaCha was given and we choose one implementation with

$$(k_0', k_1', k_2', k_3', k_4', k_5', k_6', k_7') = (z_0, z_1, z_2, z_3, z_{12}, z_{13}, z_{14}, z_{15}).$$

We do not see any significant advantages of the specific choices. Note that this additional process needs to be performed only once for a given set of nonces. Finally, we start the regular Salsa20/ChaCha with the new initial state matrix:

$$\mathbf{X}_S^{(0)} = \begin{pmatrix} c_0 & k_0' & k_1' & k_2' \\ k_3' & c_1 & v_4' & v_5' \\ t_0 & t_1 & c_2 & k_4' \\ k_5' & k_6' & k_7' & c_3 \end{pmatrix}, \quad \mathbf{X}_C^{(0)} = \begin{pmatrix} c_0 & c_1 & c_2 & c_3 \\ k_0' & k_1' & k_2' & k_3' \\ k_4' & k_5' & k_6' & k_7' \\ t_0 & t_1 & v_4' & v_5' \end{pmatrix}, \tag{17.5}$$

where

1. the predefined constants (c_0, c_1, c_2, c_3) and counters (t_0, t_1) are unchanged,
2. $(k_0', \ldots, k_7')$ is the transformed key of 256 bits,
3. the original (v_0, v_1) is replaced by the last pair nonce (v_4', v_5').

ChaCha **Adoptions**

The decision to choose between Salsa20 and ChaCha is a matter of preference. Salsa20 has the advantage of being selected as an eSTREAM finalist with more extensive security studies in the literature. On the other hand, ChaCha has slightly better security per round, a bit more elegant. It appears that ChaCha is becoming more popular, with various recent adoptions as described next.

ChaCha is the basis of the BLAKE hash function and its successor with higher speed, BLAKE2. BLAKE was a third round finalist in the National Institute for Standards and Technology's (NIST) SHA-3 Cryptographic Hash Algorithm Competition that began in 2007 and concluded in 2012 (Chang et al. 2012). For Internet security, Google has selected ChaCha together with Poly1305, a message authentication code, to replace RC4, the previous standard that has been proven vulnerable to attacks (see, for example, Vanhoef and Piessens 2015). ChaCha and Poly1305 are also used in OpenSSH.Several web browsers including Chrome and Firefox use ChaCha for TLS (Langley et al. 2015). For more usage and deployment of ChaCha and a timeline of adoptions and future support; see https://ianix.com/pub/chacha-deployment.html.

17.3   eChaCha: **Simple Effective Method to Extend** ChaCha

As noticed earlier, the difference between two successive initial state matrices in Salsa20 and ChaCha is often a difference of 1 among two blocks of 16 words because we increase the counter t_0 by 1 at each iteration. We need a counter for different initial state matrices to distinguish messages per session key. However, this feature could be the main weakness for Salsa20/ChaCha because it is difficult to make the two output blocks "look" unrelated, with a single bit difference, even after a long sequence of R rounds of transformations/scramblings. While the ARX operations used in Salsa20/ChaCha are effective for fast diffusion of bits, there is no rigorous theoretical support to "prove" that the joint distribution of the 16 word block (known as "within-block") is close to the distribution of a random sample (of size 16) from the uniform distribution over the range $0, 1, \ldots, 2^{32} - 1$. Additionally, there is no theoretical support that it can achieve (or approximate) the desired property of "independence" between successive blocks (known as "between-blocks"). In addition to uniformity, the "independence" properties (both "within-block" and "between-blocks") are important for both computer simulation and computer security applications.

Next, we propose a simple scheme to enhance the Salsa20/ChaCha.

Adding an external generator to Salsa20/ChaCha

The counter (t_0, t_1) is used as a running index $(t_0 + t_1 \times 2^{32})$ for the sequence of iterations and is also used to make sure there is no repeated use of the same initial state matrix. Since t_1 is rarely changed (increased by 1 for every 2^{32} iterations), it could be considered a waste of space. Therefore, we propose using t_1 to store the value of X_i, a sequence of random 32-bit numbers produced by an external generator, say RNG-X. Following the same notation of Salsa20/ChaCha as previously described, the diagram in Fig. 17.1 is a simple and effective extension of Salsa20/ChaCha.

We name the above proposed design eChaCha (enhanced ChaCha) and we can use a similar procedure to enhance Salsa20. For simplicity, we use the name eChaCha to include both extensions. A more formal and detailed description of the proposed eChaCha will be given later.

There are many classical pseudo-random number generators (PRNGs) that are popular in the field of computer simulation. For a recent review on the developments of classical PRNGs, see Deng and Bowman (2017). We will briefly discuss some of them later. Most of these PRNGs have nice properties such as long period length, high generating efficiency, proven theoretical equi-distribution over high dimensional space, and great empirical perfor-

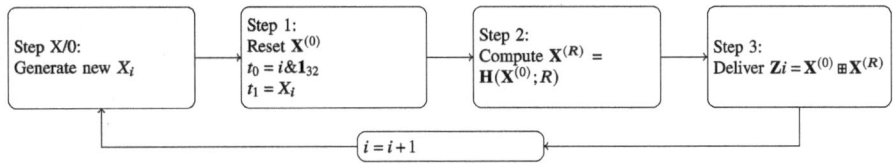

Fig. 17.1 Diagram of eChaCha method

mances. Unfortunately, most of them are linear generators and thus easily predictable with a few past observations. Adding a PRNG as in eChaCha can certainly break the linearity of the external PRNG while maintaining many of the nice properties enjoyed by classical (linear) generators. Next, we discuss additional advantages of eChaCha.

Advantages of eChaCha

There are several advantages of eChaCha:

1. ChaCha can be considered a special case of eChaCha with $X_i = i/2^{32}$ (quotient of i over 2^{32}) which will set the first 2^{32} $X_i = 0$, set next 2^{32} $X_i = 1$, set next 2^{32} $X_i = 2$, $\cdots$ and so on (a very "bad" external generator). The difference between successive initial matrices has two non-zero values: 1 at the position of "t_0" and $(X_{i+1} - X_i) \bmod 2^{32}$ at the position of "t_1". Clearly, eChaCha should have a much better diffusion property than ChaCha and it should be harder for attacks based on the powerful differential cryptanalysis. Given the near identical structure, we believe that eChaCha is at least as secure as ChaCha or Salsa20.
2. One disadvantage of eChaCha is a slight generating inefficiency when compared with the original ChaCha. The extra cost is due to the additional external generator. However, the percentage of increase in generating time is very small (around 2–5%) especially when the number of rounds, R is large. Given the fact that the classical PRNG is usually very efficient, the majority of the computing cost of eChaCha was on the part of ChaCha, especially when $R = 20$. On the positive side, adding a PRNG can allow the use of a much smaller R, say $R = 2$, and the resulting eChaCha still passes stringent empirical tests. Therefore, it is feasible to consider a smaller value of R for eChaCha to achieve better efficiency while maintaining great empirical performances.
3. With much smaller additional cost, eChaCha can allow many more nonces than ChaCha (64 bits or 2 words) or XChaCha (192 bits or 6 words). This is because lots of PRNGs that could be chosen for eChaCha have many internal states which can be initialized with a simple mixing of keys and extra nonces. eChaCha has an extra initialization for the external generator on its internal state vector which can be set with a mixture of the key and the extra nonces requested while keeping the original nonces (v_0, v_1). To eliminate

the trouble of finding a good mixer, we can adopt the initialization procedure used in XSalsa20 and XChaCha and some of the unused variates in Eq. 17.4 to initialize the internal states of the PRNG. Note that only 8 variates produced in the matrix Eq. 17.4 were used for the matrix Eq. 17.5, the remaining unused 8 words can be set to the internal state vector of the external generator. Therefore, no additional computing cost is involved for mixing the keys and IVs well and extending the possibility to allow many more nonces.

4. With a large period PRNG for X_i, the period length of $(t_0, t_1) = (i \& \mathbf{1}_{32}, X_i)$ can be much larger than 2^{64} which is the period of the counter (t_0, t_1) of ChaCha with $(t_0, t_1) = (i \& \mathbf{1}_{32}, i/2^{32})$, where $\mathbf{1}_{32}$ is a variate of 32 1's in its binary representation and $i/2^{32}$ is the quotient of the integer division. Even with a "bad" PRNG generating only constant $X_i = c$, the period length is at least 2^{32}.

5. Due to the simplicity of the counter, the original Salsa20/ChaCha has a simple "jump-ahead scheme" by changing the contents of the counter (t_0, t_1). This unique feature could be important for a parallel or random access implementation of ChaCha. For example, the user can seek directly any position in the output stream. Many PRNGs considered in the field of computer simulation also have a simple and efficient jumping ahead scheme. See, for example, Deng and Bowman (2017) for details. In that case, eChaCha ciphers will also have an efficient jumping scheme by changing the content of i and setting $t_0 = i \bmod 2^{32}$ and/or changing the starting seed of X_i and then set $t_1 = X_i$. Therefore, eChaCha can also easily perform a "random access" which is one of the main reasons for using AES in counter mode (see Stallings 2016, Chap. 7).

Design of eChaCha

In addition to the diagram for eChaCha, we can formally describe its general procedure below:

Algorithm 17.1: General procedure of eChaCha

Step X   Initialize the internal state vector of PRNG, (X_i), with a mixer of user keys and IVs which are extras in addition to the original (v_0, v_1). For simplicity, we adopt a similar procedure used in XSalsa20 and XChaCha and some of the unused variates in Eq. 17.4 to initialize the internal states of the PRNG. Let i be the running index of the iteration which can be set as $i = i_0$, where i_0 can be chosen as any number. For simplicity, we set $i = 0$.

Step 0   At the i stage, generate new X_i.

Step 1   Reset initial state matrix $\mathbf{X}^{(0)}$ in Eq. 12.1 with $t_0 = i \& \mathbf{1}_{32}, t_1 = X_i$.

Step 2   Compute R-round transformation as $\mathbf{X}^{(R)} = \mathbf{H}(\mathbf{X}^{(0)}; R)$ in Eqs. 12.10 or 12.11.

Step 3   Output sixteen 32-bit words

$$Zi = \mathbf{X}^{(0)} \boxplus \mathbf{X}^{(R)}.$$

Increase index $i = i + 1$, and repeat the whole process (0)–(3).

17.4 Variants of eChaCha

There are other interesting variants of eChaCha that can be considered and they belong to a general family of eChaCha. The first cipher, eChaCha-OFB, is motivated from the idea of improving efficiency. We propose to replace the external generator with some output feedback (OFB) which serves as the role of the (internal) PRNG. Given that no external PRNG is used, it should have near identical generation speed as the original ChaCha. The second cipher, eChaCha-XY, is motivated from the idea of further improving the distributional or diffusion property. We propose to replace both counters (t_0, t_1) with two external generators which can be viewed as the most general form of the family of eChaCha ciphers. eChaCha-XY is expected to have the ability to add more nonces, faster diffusion property, better empirical performances and more theoretical justification than the original eChaCha. Both extensions will be discussed in detail next.

17.4.1 eChaCha-OFB: Using Output Feedback in ChaCha

Another improvement under consideration is using just a single PRNG as an external generator which can clearly speed up the generating speed. The idea is the same as that behind the proposed eChaCha-OFB which uses its intermediate output as another PRNG to replace the second PRNG (Y).

There is yet another way to extend ChaCha without the necessity to add an external generator. The idea is to use an entry from the transformed matrix $\mathbf{X}^{(R)} = \mathbf{H}(\mathbf{X}^{(0)}; R)$ in Eq. 12.11 as a "generator" to replace the external generator (X_i) in eChaCha. This method is very similar to the OFB (output feedback) mode for popular block ciphers like DES and AES. We let Y_0 denote the OFB variate which is initially set from a simple mixer of user keys and nonces. We then use Y_0 taking the role of the external generator (X_i) in eChaCha to set the (upper) counter $t_1 = Y_0$ in the initial matrix $\mathbf{X}^{(0)}$. After R rounds of transformations, we use Y_0 to "randomly" select an entry from $\mathbf{X}^{(R)}$ as the next OFB (Y_i, which later will be denoted as Y_0). The random selection index can be any function of the current Y_0 but it is efficient to use $j_0 = Y_0\&1_4$. We name this extension as eChaCha-OFB and its procedure is described below:

Step 0   Create the initial state matrix $\mathbf{X}^{(0)}$ as in Eq. 12.1. Let i be the running index of the iteration. Counter (t_0, t_1) can be set as $i = i_0$, where i_0 can be chosen as any number. For simplicity, we set $i = 0$ and initialize "output feedback" (OFB) Y_0 from a simple combination of user keys or nonces.

Step 1   At the i stage, initialize state matrix $\mathbf{X}^{(0)}$ in Eq. 12.1 with counter $t_0 = i$, $t_1 = Y_0$ (OFB).

Step 2   Compute R-round transformations as $\mathbf{X}^{(R)} = \mathbf{H}(\mathbf{X}^{(0)}; R)$ in Eqs. 12.10 or 12.11. Let $j_0 = Y_0\&1_4$ be the index from the lower 4-bits of Y_0 as a (random) selection

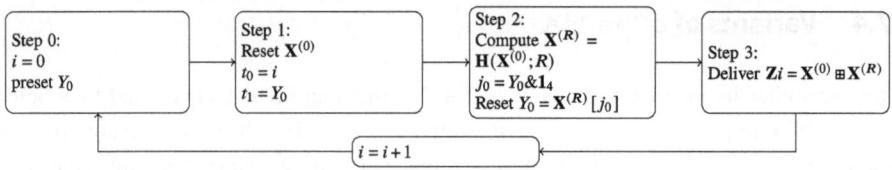

Fig. 17.2 Diagram of eChaCha-OFB method

index. Use j_0 to select an entry from $\mathbf{X}^{(R)}$ as the next OFB Y_0. That is, the next OFB is $Y_0 = \mathbf{X}^{(R)}[j_0]$.

Step 3 Output sixteen 32-bit words

$$Zi = \mathbf{X}^{(0)} \boxplus \mathbf{X}^{(R)}.$$

Increase index $i = i + 1$, and repeat the whole process (1)–(3).

The diagram for eChaCha-OFB is shown in Fig. 17.2.

eChaCha-OFB can be viewed as ChaCha running on a *combination* of count mode (CTR) (with t_0) and output feedback mode (OFB) (with t_1). Both CTR and OFB are popular modes of operations for block ciphers (Stallings 2016, Chap. 7). In the area of computer simulation, an idea similar to OFB was proposed by Bays and Durham (1976).

One clear advantage of eChaCha-OFB over eChaCha is its generating efficiency which is virtually the same as the original ChaCha because no external generator is used. On the negative side, eChaCha-OFB does not have a simple and fast "jump-ahead" for "random access" like ChaCha or eChaCha. Without an external generator, eChaCha-OFB also loses the option to greatly increase the nonce size but one can use the OFB variate (Y_0) to increase slightly the nonce size from 64 bits to 96 bits.

17.4.2 eChaCha-XY: Adding RNG-X and RNG-Y to ChaCha

The main idea of eChaCha-OFB is to use both OFB and CTR as two separate "generators" where CTR is clearly not a good one and the OFB is generated internally. It is straightforward to extend the eChaCha procedure further by adding two external PRNGs producing two sequences of X_i and Y_i, $i = 0, 1, \ldots$. This procedure simply uses counter t_0 for X_i and t_1 for Y_i. This generalized procedure is called eChaCha-XY and it can be shown in the Fig. 17.3.

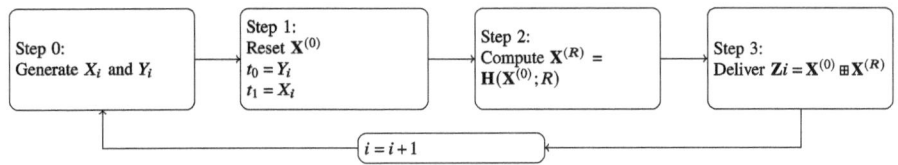

Fig. 17.3 Diagram of eChaCha-XY method

The period length of (X_i, Y_i) is the least common multiple of their periods, P_X and P_Y. Clearly, eChaCha can be considered as a special case of eChaCha-XY with $Y_i = i$. Following a similar justification for eChaCha, eChaCha-XY should have an even faster diffusion effect because the difference of two successive initial state matrices has two non-zero (large and random) values of $(X_{i+1} - X_i)$ mod 2^{32} at the "t_1" position and $(Y_{i+1} - Y_i)$ mod 2^{32} at the "t_0" position, respectively. The only drawback for eChaCha-XY is that it will (very) slightly increase the generating time.

eRabbit: Enhancements of Rabbit

18

In this chapter, we focus only on software ciphers with a special interest in Rabbit which was proposed by Boesgaard et al. (2003). Rabbit cipher has an input of a 128-bit key and an optional 64-bit initialization vector. Internally, the cipher uses a state vector and a counter vector each consisting of eight 32-bit words with an additional one-bit for storing "carry". The output of the cipher is produced from the updated state vector into four 32-bit words per iteration. The cipher's strength rests on strong mixing of its inner state and counter vectors during successive iterations. The mixing function is based on arithmetic operations available on a modern processor without the need for S-boxes or lookup tables. The mixing function uses a g-function with a squaring function, and ARX operations, that consist of logical XOR, bit-wise rotation with fixed rotation amounts, and addition modulo 2^{32}.

Rabbit cipher can be divided into several distinct modules. The first module initializes the internal counter and state vectors by expansion of the input key and optional IV with a series of non-linear transformations to achieve random and uniform distribution for the internal vectors. The second module updates the counter variables using some counter-based scheme and a carry bit so that a long period length of the counter system can yield a long period length of $2^{256} - 1$. The third module uses a non-linear g-function to mix the state vector and counter vector. This produces new G vectors which are used to update the state vector with another set of eight non-linear transformations. Finally, Rabbit produces four 32-bit words from the state vector of eight 32-bit words by various combinations of sub-words.

Rabbit stream cipher is recognized as an excellent cipher which has been known to enjoy strong security and great generating efficiency. In this chapter, simple and effective enhancements to Rabbit is also presented. We will introduce the proposed enhancements in the reverse order of the Rabbit modules: (a) update state vector with g-function (b) update counter vector (c) initialization of state/counter variables.

© The Author(s), under exclusive license to Springer Nature Switzerland AG 2025   233
L. Deng et al., *Random Number Generators for Computer Simulation and Cyber Security*, Synthesis Lectures on Mathematics & Statistics,
https://doi.org/10.1007/978-3-031-76722-7_18

According to its designer, $g(x)$ was a key component in the `Rabbit` design, squaring a 32-bit number to produce a 64-bit number, and then combining the left half and the right half of that squared number with XOR, to produce a 32-bit result. However, as pointed out first by Aumasson (2007), such a function may produce a small bias in the output of `Rabbit` resulting in a distinguisher with a very large number of runs. It is not considered a threat to `Rabbit`'s security because its complexity is significantly higher than the brute-force of the key space. We will propose an alternative function to replace the g-function.

Another possible problem for `Rabbit` is its counter updating scheme which was based on a simple "random number generator" to update its counter vector. Such counter-based schemes can ensure an extremely long period length of the count system with period length reaching $2^{256} - 1$. This updating scheme can be viewed as a simple way to produce a sequence of random numbers with constant increments (with possible carry added) at each iteration. We propose replacing this counter-update scheme with a set of eight good external generators chosen to have great distributional properties and huge period lengths.

Finally, `Rabbit`'s initialization uses a key/IV expansion scheme to fill in the state/counter vectors with an additional series of non-linear ARX operations on these vectors. However, there is no justification that `Rabbit`'s initialization scheme will ensure a nice distribution property for any choice of key/IV even after a series of non-linear ARX operations. A better strategy is to use a key/IV to initialize seed vectors for a good external generator. The external generators can then be used to fill the contents of state and counter variables. Typical good external generators will not be affected by the choice of initial seeds, and key/IV information will be quickly diluted after a short burn-in period. This procedure of using external generators to fill in initial tables/vectors was proposed first in Deng et al. (2018).

The remainder of this chapter is organized as follows. We first describe and discuss the general design of the `Rabbit` cipher in Sect. 18.1. In particular, we discuss the several modules used:(a) key/IV initialization, (b) counter vector update, and (c) state vector update. Several potential problems are discussed. In Sect. 18.3, we then propose some enhancements to these modules with some justifications provided. In particular, a simple and effective mixing method is presented which uses mutual random rotation on state and counter vectors. We also propose to include several good pseudo-random number generators for either the counter update scheme and/or key/IV initialization. Instead of expanding a key/IV to initialize state and counter variables, we can use the key/IV expansion scheme to initialize the seeds needed for the external generators and then use them to generate random numbers needed for the state/counter vector initialization. There are several good pseudo-random number generators proposed recently with extremely long period length and strong statistical properties such as the high-dimensional equi-distribution property. Section 18.4 provides a brief review of some good external generators that can be used. These generators are linear generators, and they cannot be used directly because of their weak security. Using them as a "baseline generator" inside a known secure cipher has mutual benefits. Finally, Sect. 18.4 gives concluding remarks.

18.1 Rabbit **Stream Cipher**

We begin with the description of Rabbit stream cipher proposed in Boesgaard et al. (2003). In total, there are 513 internal bits with eight 32-bit state variables, eight 32-bit counters, and one counter-carry bit. Generally speaking, these variables are updated using a series of non-linear transformations. The functions used are typical operations, known as ARX (Addition, Rotation, and XOR) operations. Rotation can ensure non-linearity; mixing/addition can hide individual values and improve the distributional property as a combination generator. The goal is to diffuse/randomize information about **K** and **IV** during the initialization step of the state variables **X** and counter variables **C**. A high-level description follows:

Algorithm 18.1: Rabbit

1. **Key/IV initialization**. These internal state variables **X** and counter variables **C** will be initialized with complicated functions of **K** and (optional) **IV**. Use a series of non-linear transformations to diffuse the information about **K** and **IV** while the state variables **X** are "random" following a "uniform" distribution.
2. **Counter vector update**. Counter variables **C** will be updated with a sequence of equations of the counter-based scheme with extra "carry" to be added in sequential update/computation.
3. **State vector update**. Produce a G-vector of eight words using specialized non-linear transformations on $g(x)$ function to mix current state vector and updated counter vectors. Update state vector, **X**, with another set of transformations on the G vector.
4. **Output extraction scheme**. Choose four extraction functions for four output variables, **S**, from the updated eight state variables, **X**.
5. Repeat (2), (3), (4) for next outputs.

Recall of Notations

To describe Rabbit stream cipher, we used the following for 32-bit integers x, y.

- $x \gg n$ denotes the right shift operator of x,
- $x \ll n$ denotes the left shift operator of x.
- $x \| y$ denotes the concatenation of the bit strings of x, y,
- $x \oplus y$ is the bit-wise logical XOR of x and y,

- $x \& y$ is the bit-wise logical AND of x and y,
- $x + y$ means $x + y \bmod 2^{32}$.
- $\lfloor x/y \rfloor$ denotes the floor/integer function of a value x/y.
- $x \ggg n$ denotes the right rotation operator $((x \gg n) \oplus (x \ll (32 - n)))$, and $x \lll n$ denotes the left rotation operator $((x \ll n) \oplus (x \gg (32 - n)))$.

18.2   Analysis of Rabbit

As mentioned, Rabbit stream cipher has been studied extensively since it was published in 2003 by Boesgaard et al. (2003). After a brief discussion of the basic properties of period length and security, we will point out some potential problems. Then possible enhancements to Rabbit will be discussed in the next section.

18.2.1 Period Length and Security Analysis

The long period length of a stream cipher is clearly an important feature to consider. According to Boesgaard et al. (2003), Rabbit counter states period length was found to be $2^{256} - 1$ due to a general property enjoyed by any counter-assisted stream ciphers. It is believed the input to the g-functions has at least the same period and a very pessimistic lower bound of 2^{215} can be guaranteed on the period of the state variable. See, for example, Boesgaard et al. (2003) and Mukherjee (2013).

The authors of Rabbit provided extensive cryptanalysis on the proposed ciphers: Partial Guessing, Algebraic Attacks, Correlation Attacks, Differential Analysis and Statistical Tests. Rabbit has been evaluated against all known attack techniques and no major weaknesses have been found. See, also Mukherjee (2013). Given that Rabbit has been selected as one of the four finalists in the eSTREAM project in ECRYPT (2005), it is considered a compact, efficient, and secure cipher.

18.2.2 Problem of Using g-Function

While there are no known effective attacks on Rabbit, Aumasson (2007) was the first to point out a potential weakness due to its use of the g-function as defined in Eq. 12.14 which is essential in the key generation scheme. It may cause a very small bias in certain keystream bits.

As mentioned by the authors of Rabbit, Boesgaard et al. (2003), the non-linear g-function is

$$g(y) = y^2 \oplus (y^2 >> 32) \bmod 2^{32},$$

for a 32-bit integer y. The goal of this g-function is to destroy the linear relations in the cipher as part of ARX operations. According to the authors, this non-linear transformation was inspired by the complex behavior of real-valued chaotic maps which can produce "random" and "unpredictable" sequences. This g-function is applied to each component of $(\mathbf{X} + \mathbf{C})$ which is the summation of the current state vector, $\mathbf{X}$ and the updated counter vector, $\mathbf{C}^*$ to produce a g vector

$$\mathbf{Y} = (\mathbf{X} + \mathbf{C}^*); \quad \mathbf{G} = g(\mathbf{Y}) = \mathbf{Y}^2 \oplus (\mathbf{Y}^2 \gg 32) \bmod 2^{32}.$$

where for a vector $\mathbf{Y}$, $\mathbf{Y}^2$ is a component-wise square.

To address the possible bias problem raised by Aumasson (2007), we will introduce an enhancement to this function.

18.2.3 Problem of Using Counter Updating Scheme

The counter update schedule proposed in Rabbit cipher is a major component that enables a quick update on the eight counter variables via Eq. 12.13. Specifically, for $j = 0, 1, 2, \ldots, 7$, update the j-counter as

$$R = c_j + 1 + \phi^*;$$
$$c_j^* = R \bmod 2^{32}; \qquad (18.1)$$
$$\phi^* = I(R \geq 2^{32}),$$

where $\phi^* = I(A)$ is an indicator function of a set A.

The designers provided no explanations for choosing the constants in Eq. 12.12. Choosing values closer to 0 or 2^{32} is not a good idea because the former will result in very few "carry" whereas the latter will set too many "carry". It is interesting to note that the given constants in Eq. 12.12 are roughly midway between 0 and 2^{32}, as seen below. This observation explains the choice.

1. $\alpha_0 = \alpha_3 = \alpha_6 = 1295307597 \approx 0.3015873015 \times 2^{32}$,
2. $\alpha_1 = \alpha_7 = 3545052371 \approx 0.8253968253 \times 2^{32}$,
3. $\alpha_2 = \alpha_5 = 886263092 \approx 0.2063492062 \times 2^{32}$,
4. $\alpha_4 = 3545052372 \approx 0.8253968256 \times 2^{32}$.
5. average is $\frac{1}{8}\sum_{i=0}^{7} 1 = 16293606089 \approx 0.4742063491 \times 2^{32}$.

To demonstrate the effect of this process, we consider just a counter system with two LCGs, proposed by Lehmer (1951), X and Y:

$$X_i = B_1 X_{i-1} + A_1 \bmod p_1$$
$$Y_i = B_2 Y_{i-1} + A_2 \bmod p_2 \qquad (18.2)$$

Other than the "carry" bit, the counter system in Rabbit is similar to simple LCGs with $B_1 = B_2 = 1$ and $p_1 = M_2 = 2^{32}$.

$$X_i = X_{i-1} + A_1 \bmod 2^{32}$$
$$Y_i = Y_{i-1} + A_2 \bmod 2^{32}. \tag{18.3}$$

For any odd numbers A_1 and A_2, the period length of X_i and Y_i will reach the maximum 2^{32}. However, for a 2-dimensional vector sequence, (X_i, X_i), will also have the same period length. Therefore, an additional "carry" bit, ϕ^* was added to the Rabbit cipher system. The "carry" bit, ϕ^*, was computed from the recent update and can be set initially to 0 or 1

$$R = X_{i-1} + A_1 + \phi^*; \quad X_i = R + \phi^* \bmod 2^{32}; \quad \phi^* = I(R \geq 2^{32})$$
$$R = Y_{i-1} + A_2 + \phi^*; \quad Y_i = R + \phi^* \bmod 2^{32}; \quad \phi^* = I(R \geq 2^{32}), \tag{18.4}$$

where the indicator function $I(R \geq 2^{32})$ can be also be computed as $\lfloor R/2^{32} \rfloor$ floor function. The idea of "carry" has been used in "add-with-carry" (AWC) and "multiply-with-carry" (MWC) by Marsaglia and his co-coauthor in Marsaglia and Zaman (1991), Marsagila (1997). In particular, MWC was given by

$$X_i = (BX_{i-1} + A_{i-1}) \bmod p, \tag{18.5}$$

and compute $A_i = \lfloor \frac{(BX_{i-1}+A_{i-1})}{p} \rfloor$ to be used as the "carry" for the next iteration and the initial A_{-1} can be set by the user. Clearly, this class of generators is similar to the equations above with "carry", A_i, as a "feedback" for the next iteration. MWC was originally proposed as a generalization of the "add-with-carry" generator of Marsaglia and Zaman (1991). The main difference with Rabbit is that the "carry" bit was used sequentially for the next generator in the system of generators used.

For Rabbit, all "generators" are "chained" in a similar fashion and the period length can greatly be increased. Other than period length, the distributional property is highly dependent on the choice of A_1 and A_2. For example, choosing small values of A_1 and A_2 will result in a long sequence without a carry produced. As a "random number generator", both AWC and MWC failed badly at a stringent test like TestU01. See, for example, L'Ecuyer and Simard (2007). Returning the Eq. 18.2 with two LCGs having two different prime modulus, p_1 and p_2. The period length of (X_i, Y_i) can reach the maximum period of the least common multiple of $(p_1 - 1, p_2 - 1)$. The only problem is that the generated values can not reach the maximum of a 32-bit integer. To overcome this, we can consider the following modification

$$X_i = B_1 X_{i-1} + A_1 \bmod p_1; \quad X_i^s = \lfloor X_i/p_1 * 2^{32} \rfloor$$
$$Y_i = B_2 Y_{i-1} + A_2 \bmod p_2; \quad Y_i^s = \lfloor Y_i/p_2 * 2^{32} \rfloor \tag{18.6}$$

The output sequence is (X_i^s, Y_i^s) which is a scaled up version of (X_i, Y_i) with the same period length.

In summary, the counter system in Rabbit used a "bad" generator to update counters and we can consider using eight good number generators to update the counter variables. More details will be provided later.

18.2.4   Problem of Using Key/IV Expansion Scheme

Compared with two previously mentioned problems, the key/IV expansion scheme problem is fairly minor. With various key/IV as inputs, most ARX operations cannot ensure uniformity/randomness of the initialized state variables $\mathbf{X}$ and counter variables $\mathbf{C}$. For Rabbit cipher, it is requested to run the "next-state function" four times after the initial key-expansion scheme, and with (optional) IV expansion another four iterations of "next-state function" are performed. However, these ARX operations cannot ensure the uniformity of these two initial vectors. Eight external generators can be used to initialize values of $\mathbf{X}$ and $\mathbf{C}$. Specifically, we can set the initial seeds of these generators using simple key/IV expansion schemes. This idea is motivated by a previously proposed design, called SAFE (Secure And Safe Encryption), proposed by Deng et al. (2018). SAFE design introduced some "good" external generators and proposed to seed the external generator with the keys/IVs and then use them to fill in the state and counter variables. We can also introduce a "good" external generator to improve the counter updating scheme of the Rabbit cipher.

Several popular stream ciphers have the same problem of key/IV initialization without using some good random number generators. HC-128 was one of the four finalists in the eSTREAM Portfolio. HC-128 is a variant of the original HC-256 proposed in Wu (2004). HC-128 has two secret tables, P and Q, each with 512 32-bit integers. At each iteration, HC-128 updates an element of one table with a non-linear feedback function of several entries in the same table and generates one 32-bit output by mixing some nonlinear functions of entries from both tables. While it is fairly efficient during the generation stage, its initialization stage requires a long clock cycle. According to Wu (2004), it is difficult to predict the exact period of the cipher, but it is estimated to be much greater than $2^{256} - 1$.

Another popular stream cipher, Salsa20/R is a family of stream ciphers introduced in Bernstein (2008b) which uses three simple operations: Addition, Rotation, XOR (ARX) for a fixed number of rounds, R. Bernstein (2008b) recommends Salsa20/20 for most applications. Salsa20/12 was submitted to the eSTREAM for Profile 1 (software based ciphers) and was chosen as one of the four finalists, see Robshaw and Billet (2008). The ChaCha cipher, proposed by Bernstein (2008a), is closely related to the Salsa20 family and aims to increase the amount of diffusion without increasing complexity. Both Salsa20 and ChaCha stream ciphers are relatively simple in their design with internal variables of 16 integer words (viewed as a 4×4 matrix) of eight words for key, two words for IV, two counter words, and four constant words stored in the initial matrix. Without using some good external generators, they relied mostly on many rounds of ARX operation hoping to achieve the randomness and uniformity distribution property for the output variables in the sequence.

We now formally present a novel and simple enhancement to `Rabbit`.

18.3   eRabbit: **Enhance** Rabbit

In this section, several simple and effective procedures to enhance modules of the `Rabbit` procedure are presented: (1) new g-function (`eRabbit`-G) (2) new counter update scheme (`eRabbit`-C) (3) new key/IV updating scheme (`eRabbit`-K), and (4) update to all three modules (`eRabbit`-A).

18.3.1 eRabbit-**G:** Rabbit **with New g-Function**

Before introducing a new g-function, we review some theory and offer vector notations that extend commonly used notations.

Statistical Theory About Uniformity

Let X and Y be random variables distributed uniformly over $0, 1, \ldots, 2^{32} - 1$ and the binary representation of X is $(D_{31} D_{30} \ldots D_1 D_0)_2$. We state the following simple facts without formal proof:

1. The binary bits of X, $D_i, i = 0, \ldots 31$ will be independent each with $P(D_i = 0) = P(D_i = 1) = 1/2$.
2. Let $r(X)$ be any $r < 32$-bit extraction from X, then random variable $r(X)$ will be distributed uniformly over $0, 1, \ldots, 2^r - 1$.
3. For any integer $0 \leq n \leq 31$, both $X \ggg n$ and $X \lll n$ will be distributed uniformly over $0, 1, \ldots, 2^{32} - 1$.
4. If X and Y are statistically independent and $r(Y)$ is any $r < 32$-bit extraction, then random variable $X \ggg r(Y)$ and $X \lll r(Y)$ will be distributed uniformly over $0, 1, \ldots, 2^{32} - 1$.
5. $Z = X + n \bmod 2^{32}$ will be uniformly distributed over $0, 1, \ldots, 2^{32} - 1$
6. If X and Y are statistically independent then for any possible distribution of Y, both $Z = X + Y \bmod 2^{32}$ and $W = X \oplus Y \bmod 2^{32}$ will be uniformly distributed over $0, 1, \ldots, 2^{32} - 1$.

It is interesting to observe that even if X and Y are not exactly uniformly distributed, the distribution of $Z = X + Y \bmod 2^{32}$ will improve the uniformity property of both X and Y. In the field of statistics, this is known as the combination generator which can mask the value of X and Y and also improve the uniformity property, see, more details in Sect. 5.4. The piling-up lemma introduced in Matsui (1993) is applied in linear cryptanalysis to construct linear approximations to the action of block ciphers. It was introduced as an analytical tool for linear cryptanalysis, not necessarily for the purpose of "improving" the uniformity distribution. The lemma states that the bias (deviation of the expected value from 1/2) of a linear Boolean function ($\oplus$)) of independent binary random variables is related to the product of the input biases.

Vectorized Notations

The following vectorized notations are useful for the procedures.

Let X and Y be two vectors of 32-bit integers, n be an integer.

1. $X \gg n$ denotes the right shift operator on each component of X, and $X \ll n$ denotes the left shift operator of on each component of X,
2. $n_1(X)$ and $n_2(X)$ denote strings extraction functions of X.
3. $X \ggg n_1(Y)$ denotes the component-wise right rotation operator on vector X with amount $n_1(Y)$, and $Y \lll n_2(X)$ denotes the component-wise left rotation operator on vector Y with amount $n_2(X)$.

We set $n_1(X) = (X >> 1)\&11111_2$ and $n_2(X) = (X >> 6)\&11111_2$ so that the range of variables in the vector $n_1(X)$ is $0., 1, \ldots, 31$ which are used to perform a (random) component-wise rotation operator on vector X. Similarly, for $Y \lll n_2(X)$, the (random) values of $n_2(X)$ will decide (random) component-wise rotation operator on vector Y.

New State Update with Mutual Random Rotations

Instead of $g(x)$ function defined in Eq. 12.14, we consider a new function:

$$G = G(X, C^*) = (X \lll n_1(C^*)) + (C^* \ggg n_2(X)) \mod 2^{32}. \qquad (18.7)$$

where $n_1(x)$ and $n_2(x)$ are bit extraction functions of 32-bit integer x as defined previously. This new function with "mutual random rotation" and "combination" operations has distinct features:

1. $n_1(C^*)$ represents a "random number of extraction" from each component of the vector C^* and $(X \lll n_1(C^*)$ is a "random number of left rotation" of X determined by $n_1(C^*)$.
2. Similarly, $n_2(X)$ represents a "random number of extraction" from X and $(C^* \ggg n_2(X))$ is a "random number of left rotation" of C^* determined by $n_2(X)$
3. The rotation will maintain the uniformity of the variation, and combination/mixing of the two vectors can further improve the uniformity. See, for example, Horton and Smith III (1949b) and Deng and George (1990).

4. Because of the "random rotation" for every entry in the two vectors, it would be impossible to apply an algebraic attack.

With new vector **G** produced, we can use the same procedure in `Rabbit` to update the state vector.

18.3.2 `eRabbit`-**C**: `Rabbit` **with New Counter Update**

The counter update schedule proposed in `Rabbit` cipher is a major component that enables a quick update on the eight counter variables via Eq. 12.13. As explained earlier, the original counter-updating scheme in Eq. 12.13 basically used a naive method of generating "random numbers" with simple increments of the constants as specified in Eq. 12.12. The "carry" bit (computed sequentially) is important so that the final period length of the counter system can reach $2^{256} - 1$. This means that parallel updates on the counter variables are infeasible.

With this simple observation, we can consider a simple enhancement, eRabbit-C, as for $j = 0, 1, 2, \ldots, 7$, update the j-counter as

$$c_j^* = c_j + \text{RNG}_j() \bmod 2^{32};$$

where $\text{RNG}_j()$ is the jth good random number generator selected. We can choose eight good generators with even longer period lengths to update the counter variables. The period length is the least common multiple of all eight generators chosen. There are so many good generators that can be used and more details will be discussed in the next section.

Several variations can be considered. For example, we can just fill in c_j^* with the j-th PRNG

$$c_j^* = \text{RNG}_j() \bmod 2^{32}$$

The problem is that we need to make sure the $\text{RNG}_j()$ is already scaled to 32-bit. The combination effect in Sect. 18.3.2 can also improve the uniformity of the counter variables. Another possibility is to introduce a "carry" bit ϕ^* as in the original `Rabbit` cipher

$$R = c_j + \text{RNG}_j(); \quad c_j^* = R + \phi^* \bmod 2^{32}; \quad \phi^* = I(R \geq 2^{32}),$$

where $I(A)$ is the indicator function. This may further increase the period length of the system, yet parallel updates remain impossible.

18.3.3 eRabbit-K: Rabbit with New Key/IV Scheme

In addition to using eight external generators with good statistical properties as in eRabbit-C, we can use additional external generators to initialize the internal state vector **X** and counter variables **C**. We use the secret user key and public IV values to initialize the starting seeds of these external generators. This step is justified because it may yield good distributional properties with any user key and IV values as inputs. To further diffuse the information about keys/IVs, a small varying number (dpending on keys/IVs) of "burn-in" variates can be discarded first.

Using a key/IV expansion scheme to seed the baseline generators has several advantages: the performance of these good number generators will not be affected by seed selection, and they produce good "random" numbers with the same period length. The information about key/IV will be quickly diffused within a few runs. The initial values of **X** and **C** are filled directly from these external generators without the need for additional rounds of ARX operations (e.g. next-state function). With this additional layer, it should be even harder to recover the key/IV.

18.3.4 eRabbit-A: Rabbit with All Components

Given the modularity of the original Rabbit cipher, we can enhance the components separately or jointly. If all three components (**G**, **C** and **K**) are used, it will be denoted as eRabbit-A.

To provide context for the formal proposal, we first provide a brief review of recent developments in classical pseudo-random number generation.

18.4   Choosing Baseline Generators for eRabbit

For the proposed design, we use eight pseudo-random number generators (PRNGs), RNG_j, $j = 1, 2, \ldots 7$, as baseline generators. Let the sequence $\{X_i, i \geq 0\}$ of each generator be defined by:

$$X_i = f(X_{i-1}, \ldots, X_{i-k}) \bmod p, \quad i \geq 0, \tag{18.8}$$

with a suitably chosen integer-valued function f of the most recent k integers in the past and k initial seeds, $X_{-k}, X_{-(k-1)}, \ldots, X_{-1}$.

Recently, many such PRNGs have been proposed mainly for computer simulation applications, and we will discuss some of the options. Since most of these classical generators are linear generators, they are not suitable for security applications.

Choice of External Baseline Generator

Let p be a prime number and $\mathbb{Z}_p = \{0, 1, \ldots, p-1\}$ be the finite field of p elements.

The MRG is a popular class of generators for computer simulation. An MRG of order k computes its next value X_i based on a k-th order linear recurrence defined as

$$X_i = \alpha_1 X_{i-1} + \alpha_2 X_{i-2} + \cdots + \alpha_k X_{i-k} \bmod p, \quad i \geq 0, \tag{18.9}$$

where $\alpha_1, \alpha_2, \ldots, \alpha_k$ are integers in $\mathbb{Z}_p$, $\alpha_k \neq 0$, and we can choose any k not-all-zero values as starting seeds, i.e., $(X_{-k}, X_{-(k-1)}, \ldots, X_{-1}) \neq (0, 0, \ldots, 0)$. It is well known that the maximum period is $p^k - 1$, which can be achieved with appropriate choices of the multipliers α_i, $i = 1, 2, \ldots, k$, see, for example, Knuth (1998). All maximum-period MRGs have the equi-distribution property up to order k, that is, over its entire period of $p^k - 1$, every t-tuple ($1 \leq t \leq k$) of integers in $\mathbb{Z}_p^t$ appears exactly the same number of times (p^{k-t}), with the exception of the all-zero tuple that appears one time less, see, for example, Lidl and Niederreiter (1994). As the order k and/or prime modulus p increases, the large period and equi-distribution properties become more and more advantageous. In addition, maximum-period MRGs of large order have desirable properties of strong statistical justifications and excellent empirical performances.

LCG: The First-Order MRG

The famous classical Linear Congruential Generator, proposed by Lehmer (1951), is a special case of the MRG with order $k = 1$, which is defined as

$$X_i = B X_{i-1} \bmod p, \tag{18.10}$$

where the multiplier B can be suitably chosen to achieve the maximum period of $p - 1$. For eRabbit design, we can consider two groups of p: (a) $p = 2^{31} - c$ for an easier "signed integer" implementation and (b) $p = 2^{32} - c$ for "unsigned integer" implementation and closer to its upper limit of $2^{32} - 1$. In Table 18.1, there are 16 different primes listed in each of the two groups and 8 multipliers B for each prime p. Choosing LCGs from Table 18.1 for the required 8 baseline generators is quite straightforward. To maximize the period length, we should choose 8 different p (out of 32 listed) and then choose any of the eight B for each p in Table 18.1. In total, there are $\binom{32}{8} \times 8^8 \approx 1.764677911 \times 10^{14}$ selections. To further enhance the security or complexity, random seed selection schemes can be functions of the input key/IV.

The LCG is efficient and simple, but it is highly insecure if used alone or directly. However, LCG has a much better distributional property than the additive generator used in the original Rabbit design. After a series of ARX operations in eRabbit design, it should be just as difficult to recover the key/IV from the output sequence.

Table 18.1 List of maximum-period LCG with multiplier $B \leq 2^{16}$, modulus p

$p = 2^{31} - c$	c	B							
2147483647	1	32770	32775	32776	32779	32790	32791	32794	32795
2147483629	19	32769	32771	32773	32776	32777	32778	32784	32785
2147483587	61	32769	32775	32777	32782	32785	32796	32798	32801
2147483579	69	32771	32774	32775	32776	32777	32780	32782	32784
2147483563	85	32770	32784	32785	32790	32791	32793	32803	32808
2147483549	99	32769	32770	32771	32774	32775	32777	32779	32781
2147483543	105	32772	32773	32774	32775	32779	32783	32784	32785
2147483497	151	32769	32772	32774	32779	32781	32785	32790	32792
2147483489	159	32770	32772	32774	32777	32778	32779	32780	32781
2147483477	171	32770	32775	32778	32779	32780	32782	32783	32784
2147483423	225	32772	32775	32776	32780	32784	32787	32791	32793
2147483399	249	32769	32771	32772	32773	32774	32775	32777	32781
2147483353	295	32772	32773	32779	32787	32791	32792	32799	32803
2147483323	325	32771	32772	32774	32775	32779	32780	32781	32785
2147483269	379	32773	32776	32781	32785	32786	32788	32792	32796
2147483249	399	32769	32770	32771	32773	32774	32775	32778	32781
$p = 2^{32} - c$	c	B							
4294967291	5	32773	32775	32776	32777	32785	32787	32788	32790
4294967279	17	32771	32772	32773	32775	32780	32781	32782	32783
4294967231	65	32774	32778	32780	32786	32787	32788	32793	32796
4294967197	99	32769	32778	32782	32783	32784	32786	32790	32792
4294967189	107	32773	32774	32775	32777	32779	32780	32783	32784
4294967161	135	32769	32772	32778	32784	32790	32791	32804	32812
4294967143	153	32769	32770	32771	32773	32776	32777	32781	32784
4294967111	185	32769	32774	32780	32782	32783	32785	32791	32793
4294967087	209	32770	32771	32775	32777	32779	32782	32785	32786
4294967029	267	32773	32778	32783	32795	32796	32798	32799	32800
4294966997	299	32769	32771	32773	32775	32780	32782	32783	32784
4294966981	315	32776	32780	32786	32790	32792	32794	32796	32799
4294966943	353	32771	32773	32774	32776	32778	32779	32780	32781
4294966927	369	32772	32776	32778	32781	32783	32785	32791	32794
4294966909	387	32769	32777	32778	32779	32787	32794	32795	32798
4294966877	419	32769	32770	32773	32776	32778	32779	32781	32786

One more problem of using an LCG from Table 18.1 is that X_i can only range from 1 to $p - 1$, not very close to its upper bound $2^{32} - 1$. One simple solution is to use X_i^s from the following

$$X_i = B X_{i-1} \bmod p, \quad X_i^s = \lfloor X_i / p * 2^{32} \rfloor. \tag{18.11}$$

The main advantage of choosing LCGs as baseline generators is their simplicity and efficiency. The period length for the eight LCGs is the least common multiple (LCM) of the period length of each baseline LCG ($p_i - 1, i = 1, 2, \ldots, 8$) and depends on the specific selection of the modulus p_i. For example, choosing the first eight LCGs with $p = 2^{31} - c$ is $7.419360766 \times 10^{67}$ and with $p = 2^{32} - c$ is $1.689310054 \times 10^{69}$.

In total, there are eight starting seeds to be initialized, the same size as the input of four words for each key and IV. We can also use any key/IV expansion scheme to initialize these seeds with non-zero values. As mentioned, any non-zero seed of LCG will yield the same period length.

DX-k: An Efficient Class of Maximum-Period MRGs

For MRG, there are several proposals to yield a more efficient generator with two nonzero terms in Eq. (18.9). See, for example, Grube (1973), L'Ecuyer et al. (1993). Deng and Xu (2003), Deng (2005) proposed a general class of efficient MRGs, called DX-k generators, with special cases including DX-k-1 generators

$$X_i = (X_{i-t} + B X_{i-k}) \bmod p, \quad X_i^s = \lfloor (X_i + 1/2)/p * 2^{32} \rfloor \tag{18.12}$$

and DX-k-2 generators

$$X_i = B(X_{i-t} + X_{i-k}) \bmod p, \quad X_i^s = \lfloor (X_i + 1/2)/p * 2^{32} \rfloor. \tag{18.13}$$

Here B is a coefficient suitably chosen to achieve the full period of $p^k - 1$. Note also we used $(X + 1/2)/p * 2^{32}$ because X values are ranging $X = 0, \cdots, p - 1$. See, for example, Deng and Xu (2003).

To further improve the period length and distributional property of LCGs, we can choose these eight baseline generators in eRabbit from DX generators. A list of various DX-1/DX-2 with the order $k = 2$, $k = 5$, and $k = 10$ are listed in the end of this chapter in Tables 18.2, 18.3 and 18.4. Like the previous discussion for LCG, we can consider two groups of p: (a) $p = 2^{31} - c$ for an easier "signed integer" implementation and (b) $p = 2^{32} - c$ for "unsigned integer" implementation and closer to its upper limit of $2^{32} - 1$. In each of these tables, there are 16 different primes listed in each of the two groups and 8 multipliers B for each prime p. Choosing DX for the required 8 baseline generators is quite straightforward. To maximize the period length, we should choose 8 different p (out of 32 listed) and then choose any of the eight B for each p in DX-k-s ($s = 1, s = 2$). Similar to previous discussion of choosing various LCGs, within each table, there are $\binom{32}{8} \times 8^8 \approx 1.764677911 \times 10^{14}$ selections of various DX generators. Furthermore, we can choose generators across various tables for such section and their seeds scheme can be set as functions of input key/IV.

Table 18.2 List of DX-2-s ($s = 1, 2$) with multiplier $B \leq 2^{16}$, modulus p

$p = 2^{31} - c$	c	$s = 1$				$s = 2$			
2147483647	1	32747	32732	32728	32724	32750	32747	32744	32732
2147483629	19	32748	32747	32746	32740	32748	32747	32735	32717
2147483587	61	32745	32732	32707	32698	32746	32736	32726	32707
2147483579	69	32735	32733	32731	32725	32750	32743	32730	32725
2147483563	85	32743	32735	32733	32715	32733	32730	32715	32708
2147483549	99	32746	32745	32740	32728	32731	32729	32714	32709
2147483543	105	32738	32735	32728	32724	32744	32743	32738	32737
2147483497	151	32737	32734	32728	32700	32737	32734	32717	32716
2147483489	159	32750	32749	32746	32741	32741	32736	32723	32678
2147483477	171	32742	32741	32733	32709	32745	32743	32742	32730
2147483423	225	32726	32707	32704	32686	32739	32734	32728	32727
2147483399	249	32749	32733	32730	32725	32741	32724	32700	32699
2147483353	295	32745	32719	32718	32717	32747	32745	32736	32735
2147483323	325	32749	32744	32728	32727	32749	32746	32743	32730
2147483269	379	32750	32743	32719	32717	32717	32712	32710	32709
2147483249	399	32741	32739	32727	32710	32743	32714	32705	32700
$p = 2^{32} - c$	c	$s = 1$				$s = 2$			
4294967291	5	32749	32725	32722	32714	32744	32735	32726	32698
4294967279	17	32748	32736	32705	32704	32739	32733	32728	32713
4294967231	65	32737	32735	32734	32729	32737	32736	32727	32720
4294967197	99	32750	32749	32745	32741	32750	32749	32740	32739
4294967189	107	32740	32731	32730	32720	32748	32747	32740	32737
4294967161	135	32736	32733	32727	32726	32746	32743	32738	32736
4294967143	153	32747	32741	32740	32736	32747	32734	32727	32724
4294967111	185	32738	32735	32734	32722	32749	32732	32726	32725
4294967087	209	32749	32747	32737	32730	32744	32741	32738	32731
4294967029	267	32731	32716	32710	32709	32737	32734	32726	32719
4294966997	299	32747	32741	32733	32721	32749	32747	32728	32719
4294966981	315	32735	32731	32720	32716	32745	32740	32735	32726
4294966943	353	32732	32711	32707	32706	32747	32739	32738	32716
4294966927	369	32738	32731	32728	32727	32744	32738	32731	32725
4294966909	387	32750	32745	32737	32733	32745	32737	32710	32699
4294966877	419	32746	32744	32743	32711	32744	32743	32730	32727

Table 18.3 List of DX-5-s ($s = 1, 2$) with multiplier $B \leq 2^{16}$, modulus p

$p = 2^{31} - c$	c	$s = 1$				$s = 2$			
2147483647	1	32730	32719	32710	32703	32738	32712	32680	32659
2147483629	19	32736	32717	32714	32705	32732	32680	32650	32649
2147483587	61	32742	32737	32734	32733	32747	32734	32733	32694
2147483579	69	32738	32736	32721	32716	32748	32745	32738	32732
2147483563	85	32723	32680	32674	32641	32748	32746	32734	32701
2147483549	99	32746	32723	32714	32711	32730	32728	32709	32708
2147483543	105	32741	32733	32722	32720	32748	32733	32720	32706
2147483497	151	32734	32711	32683	32649	32733	32711	32688	32675
2147483489	159	32750	32737	32719	32718	32743	32732	32729	32719
2147483477	171	32738	32731	32703	32702	32744	32739	32729	32697
2147483423	225	32736	32731	32724	32711	32750	32743	32731	32729
2147483399	249	32747	32745	32742	32729	32747	32701	32645	32599
2147483353	295	32719	32708	32704	32623	32702	32627	32602	32578
2147483323	325	32737	32719	32718	32713	32747	32726	32707	32702
2147483269	379	32719	32708	32707	32687	32726	32708	32689	32687

$p = 2^{32} - c$	c	$s = 1$				$s = 2$			
4294967291	5	32748	32716	32715	32701	32736	32707	32696	32691
4294967279	17	32750	32744	32716	32675	32750	32726	32719	32717
4294967231	65	32740	32702	32676	32627	32712	32680	32679	32659
4294967197	99	32750	32693	32681	32674	32749	32741	32739	32723
4294967189	107	32709	32708	32698	32693	32747	32737	32714	32709
4294967161	135	32739	32738	32722	32697	32736	32726	32682	32643
4294967143	153	32715	32706	32699	32694	32738	32735	32732	32711
4294967111	185	32741	32716	32701	32699	32746	32736	32727	32697
4294967087	209	32746	32745	32733	32725	32736	32734	32728	32724
4294967029	267	32713	32677	32664	32650	32736	32709	32692	32689
4294966997	299	32715	32706	32696	32695	32749	32738	32728	32716
4294966981	315	32745	32731	32716	32713	32726	32720	32670	32655
4294966943	353	32750	32734	32713	32687	32750	32736	32662	32643
4294966927	369	32750	32742	32739	32736	32749	32742	32724	32710
4294966909	387	32730	32727	32726	32675	32750	32746	32730	32727
4294966877	419	32726	32715	32654	32638	32744	32715	32711	32698

Table 18.4 List of DX-10-s ($s = 1, 2$) with multiplier $B \leq 2^{16}$, modulus p

$p = 2^{31} - c$	c	$s = 1$				$s = 2$			
2147483647	1	32725	32724	32713	32689	32729	32689	32683	32656
2147483629	19	32733	32628	32627	32624	32747	32726	32664	32595
2147483587	61	32662	32649	32634	32627	32683	32649	32641	32569
2147483579	69	32725	32715	32713	32711	32743	32618	32571	32560
2147483563	85	32738	32707	32675	32673	32717	32646	32645	32606
2147483549	99	32664	32592	32509	32503	32635	32615	32590	32491
2147483543	105	32719	32701	32695	32694	32679	32673	32634	32625
2147483497	151	32590	32575	32477	32425	32716	32696	32634	32623
2147483489	159	32557	32514	32457	32414	32668	32585	32578	32576
2147483477	171	32740	32730	32702	32638	32668	32608	32561	32551
2147483423	225	32741	32734	32658	32648	32722	32665	32662	32587
2147483399	249	32711	32682	32657	32648	32744	32731	32684	32656
2147483353	295	32747	32709	32704	32597	32609	32578	32516	32500
2147483323	325	32746	32637	32630	32610	32746	32678	32539	32538
2147483269	379	32717	32674	32640	32611	32674	32525	32514	32496
2147483249	399	32689	32518	32480	32201	32702	32693	32618	32606
$p = 2^{32} - c$	c	$s = 1$				$s = 2$			
4294967291	5	32726	32719	32603	32525	32622	32607	32501	32498
4294967279	17	32705	32703	32509	32488	32713	32647	32604	32486
4294967231	65	32694	32568	32559	32500	32686	32655	32623	32595
4294967197	99	32722	32710	32693	32675	32708	32660	32653	32643
4294967189	107	32731	32674	32624	32603	32737	32635	32580	32554
4294967161	135	32736	32727	32726	32722	32743	32736	32726	32573
4294967143	153	32705	32698	32687	32630	32710	32701	32698	32648
4294967111	185	32702	32688	32662	32644	32745	32734	32717	32595
4294967087	209	32729	32716	32682	32662	32743	32711	32709	32680
4294967029	267	32552	32425	32368	32365	32652	32609	32538	32513
4294966997	299	32745	32741	32738	32699	32720	32682	32668	32641
4294966981	315	32745	32716	32666	32502	32736	32701	32680	32641
4294966943	353	32675	32632	32602	32566	32668	32605	32528	32470
4294966927	369	32717	32632	32621	32596	32738	32715	32628	32527
4294966909	387	32733	32730	32726	32693	32745	32682	32601	32515
4294966877	419	32745	32695	32671	32669	32745	32691	32651	32612

Table 18.5 List of period lengths for eRabbit-C and eRabbit-A with modulus $p = 2^{31} - c/p = 2^{32} - c$ and $k = 2$, $k = 5$, $k = 10$

	Modulus p	$k = 2$	$k = 5$	$k = 10$
eRabbit-C	$2^{31} - c$	1.4519×10^{133}	2.3331×10^{363}	1.2269×10^{719}
	$2^{32} - c$	2.6389×10^{132}	1.5085×10^{370}	1.7428×10^{730}
eRabbit-A	$2^{31} - c$	6.9747×10^{148}	1.3320×10^{409}	1.5636×10^{806}
	$2^{32} - c$	5.0706×10^{149}	1.0021×10^{417}	1.3008×10^{821}

Example 18.1 (*Period lengths for different setup*)

First, we select the smallest order k for DX generators, which is $k = 2$ with modulus $p = 2^{31} - c$ or $p = 2^{32} - c$. Alternatively, we have the option to choose DX-k generators with a larger order, specifically $k = 5$, resulting in a longer period length for the baseline generators. Moreover, opting for DX-k generators with an even larger order, $k = 10$, will further extend the period length of the baseline generators. These choices and corresponding tables can be found in Tables 18.2, 18.3, and 18.4 at the end of this chapter.

For any order k, the period length of each DX-k-s ($s = 1, 2$) generator is $p^k - 1$, with modulus p chosen as $p = 2^{31} - c$ or $p = 2^{32} - c$. The period length of a set of DX-k-s ($s = 1, 2$) generators is the LCM of the period lengths of each baseline generator ($p_i^k - 1$, for $i = 1, 2, \ldots, S$) and depends on the specific set S of modulus values p_i selected. The size of S for eRabbit-C is 8, whereas the size of S for eRabbit-A is 9 (eight plus one extra PRNG). The typical period lengths for various selections are summarized in Table 18.5.

In general, there are $k \times S$ ($S = 8$ or $S = 9$) starting seeds to be initialized which is more than input key/IV with $k \geq 2$. We can use any key/IV expansion scheme to initialize these seeds. As mentioned, any seeds (not all zero) will yield the same period length. Note that the period of various eRabbit ciphers ranges from 2.6389×10^{132} ($k = 2$ and $p = 2^{32} - c$ for eRabbit-C) to 1.3008×10^{821} ($k = 10$ and $p = 2^{32} - c$ for eRabbit-A). These period lengths are much longer than the period length of Rabbit with $2^{256} - 1 \approx 1.1579 \times 10^{77}$. We should note that much larger k for DX-k generators can be found using the algorithm proposed by Deng (2005).

Concluding Remarks

We have proposed several variants of eRabbit, simple but effective enhancements of the popular Rabbit stream ciphers. We summarize the main advantages of various variants of eRabbit:

1. eRabbit-G uses a random mutual rotation to replace the original g-function which can reduce the bias bit and increase the security property. eRabbit-G is recommended if one would like to maintain the maximum similarity to the original design while addressing the possible bias bit problem pointed out by Aumasson (2007). With random mutual rotations between state and counter variables, eRabbit-G is expected to remain secure without the issue of bias. With eRabbit-G, the state counter variables should have a better distribution property. eRabbit-G can maintain or perhaps increase the security property with minor modifications to the original design of Rabbit cipher.
2. eRabbit-C uses eight good external generators as baseline generators to constantly update the eight counter variables to improve the uniformity and randomness of the counter variables. Baseline generators can be chosen to achieve a much longer period length of the counter update module.
3. eRabbit-K uses one extra external generator as a baseline generator with initial seed vector from a simple key/IV expansion scheme. Since baseline generators are not affected by the choice of seed, we expect resultant state and counter vectors to follow a desired uniformity property.
4. Variants of eRabbit should be as secure as Rabbit. For a maximum enhancement to Rabbit, we can consider eRabbit-A which includes all three enhancements to Rabbit. Finally, by introducing external generators such as the DX-k generators, the period length of eRabbit can be greatly increased.

References

Agrawal M, Kayal N, Saxena N (2004) PRIMES is in P. Annals of mathematics 160(2):781–793

Alanen J, Knuth DE (1964) Tables of finite fields. Sankhyā: The Indian Journal of Statistics, Series A pp 305–328

Anderson NH, Titterington DM (1993) Cross-correlation between simultaneously generated sequences of pseudo-random uniform deviates. Statistics and Computing 3:61–65

Aumasson JP (2007) On a bias of rabbit. In: State of the Art of Stream Ciphers Workshop (SASC 2007), eSTREAM, ECRYPT Stream Cipher Project, Report, Citeseer, vol 29

Aumasson JP, Fischer S, Khazaei S, Meier W, Rechberger C (2008) New features of Latin dances: analysis of Salsa, ChaCha and Rumba. In: Fast Software Encryption, Springer, pp 470–488

Babbage S, Dodd M (2008) The mickey stream ciphers. In: New Stream Cipher Designs: The eSTREAM Finalists, Springer, pp 191–209

Barker E, Kelsey J (2012) Recommendation for random number generation using deterministic random bit generators. NIST Special Publication pp 800–90A

Bays C, Durham SD (1976) Improving a poor random number generator. ACM Transactions on Mathematical Software (TOMS) 2(1):59–64

Berbain C, Billet O, Canteaut A, Courtois N, Gilbert H, Goubin L, Gouget A, Granboulan L, Lauradoux C, Minier M, et al (2008) Sosemanuk, a fast software-oriented stream cipher. New Stream Cipher Designs: The eSTREAM Finalists pp 98–118

Bernstein D (2008a) ChaCha, a variant of Salsa20. In: In Workshop Record of SASC 2008: The State of the Art of Stream Ciphers, URL https://cr.yp.to/papers.html#chcha

Bernstein D (2008b) The Salsa20 family of stream ciphers. In: Robshaw M, Billet O (eds) New stream cipher designs: the eSTREAM finalists, LNCS, vol 4986, pp 84–97, URL http://cr.yp.to/papers.html#salsafamily

Bernstein DJ (2011) Extending the salsa20 nonce. In: Workshop record of Symmetric Key Encryption Workshop, vol 2011

Blum L, Blum M, Shub M (1986) A simple unpredictable pseudo-random number generator. SIAM Journal on Computing 15:364–383

Boesgaard M, Vesterager M, Pedersen T, Christiansen J, Scavenius O (2003) Rabbit: A new high-performance stream cipher. In: Fast Software Encryption: 10th International Workshop, FSE 2003, Lund, Sweden, February 24-26, 2003. Revised Papers 10, Springer, pp 307–329

© The Editor(s) (if applicable) and The Author(s), under exclusive license to Springer Nature Switzerland AG 2025
L. Deng et al., *Random Number Generators for Computer Simulation and Cyber Security*, Synthesis Lectures on Mathematics & Statistics, https://doi.org/10.1007/978-3-031-76722-7

Bokhari MU, Alam S, Masoodi FS (2012) Cryptanalysis techniques for stream cipher: a survey. International Journal of Computer Applications 60(9)

Boyar J (1989) Inferring sequences produced by pseudo-random number generators. Journal of the ACM (JACM) 36(1):129–141

Brown M, Solomon H, et al (1979) On combining pseudorandom number generators. The Annals of Statistics 7(3):691–695

Chang Sj, Perlner R, Burr WE, Turan MS, Kelsey JM, Paul S, Bassham LE (2012) Third-round report of the sha-3 cryptographic hash algorithm competition. NIST Interagency Report 7896:121

Choudhuri AR, Maitra S (2016a) Differential cryptanalysis of Salsa and ChaCha - an evaluation with a hybrid model. IACR Cryptology ePrint Archive p 377

Choudhuri AR, Maitra S (2016b) Significantly improved multi-bit differentials for reduced round Salsa and ChaCha. In: IACR Transactions on Symmetric Cryptology, pp 261–287

Coddington P (1994) Analysis of random number generators using monte carlo simulation. International Journal of Modern Physics C5:547

Cohen H (1993) A course in computational algebraic number theory, vol 138. Springer-Verlag

Coppersmith D, Krawczyk H, Yishay M (1994) The shrinking generator. Advances in Cryptology–CRYPTO 93 (LNCS 773) pp 22–39

Couture R, L'Ecuyer P (1997) Distribution properties of multiply-with-carry random number generators. Mathematics of Computation 66:591–607

Crandall R, Pomerance C (2006) Prime numbers: a computational perspective, vol 182. Springer Science & Business Media

Crowley P (2005) Truncated differential cryptanalysis of five rounds of Salsa20. In: Workshop Record of SASC 2006: Stream Ciphers Revisited, eSTREAM technical report 2005/073

Damgard I, Landrock P, Pomerance C (1993) Average case error estimates for the strong probable prime test. Mathematics of Computation 61(203):177–194

De Canniere C, Preneel B (2008) Trivium. In: New Stream Cipher Designs: The eSTREAM Finalists, Springer, pp 244–266

Deng LY (2004) Generalized mersenne prime number and its application to random number generation. In: Monte Carlo and Quasi-Monte Carlo Methods 2002: Proceedings of a Conference held at the National University of Singapore, Republic of Singapore, November 25–28, 2002, Springer, pp 167–180

Deng LY (2005) Efficient and portable multiple recursive generators of large order. ACM Transactions on Modeling and Computer Simulation (TOMACS) 15(1):1–13

Deng LY (2008) Issues on computer search for large order multiple recursive generators. In: Monte Carlo and Quasi-Monte Carlo Methods 2006, Springer, pp 251–261

Deng LY (2016) Recent developments on pseudo-random number generators and their theoretical justifications. Journal of the Chinese Statistical Association 54:154–179

Deng LY, Bowman D (2017) Developments in pseudo-random number generators. WIRES

Deng LY, Chu YC (1991) Combining random number generators. In: Proceedings of the 23rd Conference on Winter Simulation, IEEE Computer Society, Washington, DC, USA, WSC '91, pp 1043–1046, URL http://dl.acm.org/citation.cfm?id=304238.304413

Deng LY, George EO (1990) Generation of uniform variates from several nearly uniformly distributed variables. Communications in Statistics-Simulation and Computation 19(1):145–154

Deng LY, Lin DK (2000) Random number generation for the new century. The American Statistician 54(2):145–150

Deng LY, Shiau JJH (2015) Uniform random numbers. Wiley StatsRef: Statistics Reference Online pp 1–14

Deng LY, Xu H (2003) A system of high-dimensional, efficient, long-cycle and portable uniform random number generators. ACM Transactions on Modeling and Computer Simulation (TOMACS) 13(4):299–309

Deng LY, Rousseau C, Yuan Y (1992) Generalized lehmer-tausworthe random number generators. In: Proceedings of the 30th annual Southeast regional conference, pp 108–115

Deng LY, Chan KH, Yuan Y (1994) Random number generators for multiprocessor systems. International Journal of Modeling & Simulation 14(4):185–191

Deng LY, Lin DK, Wang J, Yuan Y (1997) Statistical justification of combination generators. Statistica Sinica pp 993–1003

Deng LY, Guo R, Lin DK, Bai F (2008a) Improving random number generators in the monte carlo simulations via twisting and combining. Computer Physics Communications 178:401–408

Deng LY, Li H, Shiau JJH, Tsai GH (2008b) Design and implementation of efficient and portable multiple recursive generators with few zero coefficients. In: Monte Carlo and Quasi-Monte Carlo Methods 2006, Springer, pp 263–273

Deng LY, Li H, Shiau JJH (2009a) Scalable parallel multiple recursive generators of large order. Parallel Computing 35(1):29–37

Deng LY, Shiau JJH, Tsai GH (2009b) Parallel random number generators based on large order multiple recursive generators. In: Monte Carlo and Quasi-Monte Carlo Methods 2008, Springer, pp 289–296

Deng LY, Shiau JJH, Lu HHS (2012a) Efficient computer search of large-order multiple recursive pseudo-random number generators. Journal of Computational and Applied Mathematics 236(13):3228–3237

Deng LY, Shiau JJH, Lu HHS (2012b) Large-order multiple recursive generators with modulus $2^{31} - 1$. INFORMS Journal on Computing 24(4):636–647

Deng LY, Shiau JJH, Lu HHS, Bowman D (2018) Secure and fast encryption (safe) with classical random number generators. ACM Transactions on Mathematical Software (TOMS) 44(4)

Deng LY, Bowman D, Yang CC, Lu HHS (2021) Extending rc4 to construct secure random number generators. In: 2021 Annual Modeling and Simulation Conference (ANNSIM), IEEE, pp 1–12

Deng LY, Winter BR, Shiau JJH, Lu HHS, Kumar N, Yang CC (2023) Parallelizable efficient large order multiple recursive generators. Parallel Computing p 103036

ECRYPT N (2005) estream-the ecrypt stream cipher project. The eSTREAM Portfolio Page Available online: http://www.ecrypteuorg/stream/. Accessed on 25 November 2020

Entacher K, Schell T, Uhl A (2002) Efficient lattice assessment for lcg and glp parameter searches. Mathematics of Computation 71(239):1231–1242

Entacher K, Schell T, Uhl A (2005) Bad lattice points. Computing 75(4):281–295

Ferrenberg AM, Landau D, Wong YJ (1992) Monte carlo simulations: Hidden errors from "good" random number generators. Physical Review Letters 69(23):3382

Fincke U, Pohst M (1985) Improved methods for calculating vectors of short length in a lattice, including a complexity analysis. Mathematics of computation 44(170):463–471

Fluhrer S, McGrew D (2000) Statistical analysis of the alleged rc4 keystream generator. in proceedings fast software encryption 2000. Lecture Notes in Computer Science 1978:19–30

Franklin JN (1964) Equidistribution of matrix-power residues modulo one. Mathematics of Computation 18(88):560–568

Frederickson P, Hiromoto R, Jordan TL, Smith B, Warnock T (1984) Pseudo-random trees in monte carlo. Parallel Computing 1(2):175–180

Geffe P (1973) How to protect data with ciphers that are really hard to break. Electronics 46:99–101

Gentle JE (2003) Random Number Generations and Monte Carlo Methods, 2nd edn. Springer-Verlag

Golic JD (2001) Correlation analysis of the shrinking generator. Advances in Cryptology–CRYPTO 2001 pp 440–457

Golomb SW (1982) Shift register sequences. Aegean Park Press

Gong G, Gupta KC, Hell M, Nawaz Y (2005) Towards a general RC4-like keystream generator. In: CISC 2005, Lecture Notes in Computer Science, vol 3822, Springer, pp 162–174

Good I (1953) The serial test for sampling numbers and other tests for randomness. Mathematical Proceedings of the Cambridge Philosophical Society 49(2):276–284

Goresky M, Klapper A (2003) Efficient multiply-with-carry random number generators with maximal period. ACM Transactions on Modeling and Computer Simulation 13(4):1–12

Greenwood RE (1955) Coupon collector's test for random digits. Mathematical Tables and Other Aids to Computation pp 1–5

Grothe H (1987) Matrix generators for pseudo-random vector generation. Statistische Hefte 28(1):233–238

Grube A (1973) Mehrfach rekursiv-erzeugte pseudo-zufallszahlen. ZAMM-Journal of Applied Mathematics and Mechanics/Zeitschrift für Angewandte Mathematik und Mechanik 53(12):T223–T225

Günther C (1988) Alternating step generators controlled by de bruijn sequences. Advances in Cryptology - EUROCRYPT 87 (LNCS 304):5–14

Haramoto H, Matsumoto M, L'Ecuyer P (2008a) A fast jump ahead algorithm for linear recurrences in a polynomial space. Proceedings of the 5th International Conference on Sequences and their Applications, SETA 08 pp 290–298

Haramoto H, Matsumoto M, Panneton P, L'Ecuyer P (2008b) Efficient jump ahead for f_2-linear random number generators. INFORMS J Comput 20:385–390

Hastad J, Shamir A (1985) The cryptographic security of truncated linearly related variables. In: Proceedings of the seventeenth annual ACM symposium on Theory of computing, pp 356–362

Hell M, Johansson T, Meier W (2007) Grain: a stream cipher for constrained environments. International journal of wireless and mobile computing 2(1):86–93

Hellekalek P (1998) Good random number generators are (not so) easy to find. Mathematics and Computers in Simulation 46(5):485–505

Hong J, Sarkar P (2005) New applications of time memory data tradeoffs. In: Advances in Cryptology-ASIACRYPT 2005: 11th International Conference on the Theory and Application of Cryptology and Information Security, Chennai, India, December 4-8, 2005. Proceedings 11, Springer, pp 353–372

Horton HB (1948) A method for obtaining random numbers. The Annals of Mathematical Statistics 19(1):81–85

Horton HB, Smith III RT (1949a) A direct method for producing random digits in any number system. The Annals of Mathematical Statistics pp 82–90

Horton HB, Smith III RT (1949b) A direct method for producing random digits in any number system. The Annals of Mathematical Statistics pp 82–90

Ishiguro T, Kiyomoto S, Miyake Y (2011) Latin dances revisited: new analytic results of salsa20 and chacha. In: Information and Communications Security: 13th International Conference, ICICS 2011, Beijing, China, November 23-26, 2011. Proceedings 13, Springer, pp 255–266

James F (1994) Ranlux: A fortran implementation of the high-quality pseudorandom number generator of lüscher. Computer Physics Communications 79(1):111–114

Johansson T, Jönsson F (2000) Fast correlation attacks through reconstruction of linear polynomials. Advances in Cryptology - CRYPTO 2000 pp 300–315

Kao C, Tang HC (1997) Systematic searches for good multiple recursive random number generators. Computers & Operations Research 24(10):899–905

Kendall MG, Babington-Smith B (1939) Second paper on random sampling numbers. Supplement to the Journal of the Royal Statistical Society 6(1):51–61

Kerckhoffs A (1883) La cryptographie militaire, ou, Des chiffres usités en temps de guerre: avec un nouveau procédé de déchiffrement applicable aux systèmes à double clef. Librairie militaire de L. Baudoin

Kirkpatrick S, Stoll EP (1981) A very fast shift-register sequence random number generator. Journal of Computational Physics 40(2):517–526

Knuth DE (1998) The art of computer programming, vol 2: seminumerical algorithms, 3rd edn. Addison-Wesley

Langley A, Chang W, Mavrogiannopoulos N, Strombergson J, Josefsson S (2015) ChaCha20-Poly1305 cipher suites for Transport Layer Security (tls). RFC 7905, Internet Research Task Force (IRTF) URL https://tools.ietf.org/html/rfc7905

Learmonth G, Lewis P (1973) Statistical tests of some widely used and recently proposed uniform random number generators. In: Kennedy WJ (ed) Computer Science and Statistics: 7th Annual Symposium on the Interface, Statistical Laboratory, Iowa State University, Ames, Iowa, pp 163–171

L'Ecuyer P (1990) Random numbers for simulation. Communications of the ACM 33(10):85–97

L'Ecuyer P (1996) Combined multiple recursive random number generators. Operations Research 44(5):816–822

L'Ecuyer P (1997) Bad lattice structures for vectors of non successive values produced by some linear recurrences. INFORMS Journal on Computing 9(1):57–60

L'Ecuyer P (1999) Good parameters and implementations for combined multiple recursive random number generators. Operations Research 47(1):159–164

L'Ecuyer P, Blouin F (1988) Linear congruential generators of order $k > 1$. In: Winter Simulation Conference: Proceedings of the 20th conference on Winter simulation, pp 432–439

L'Ecuyer P, Couture R (1997) An implementation of the lattice and spectral tests for multiple recursive linear random number generators. INFORMS Journal on Computing 9(2):206–217

L'Ecuyer P, Simard R (2007) Testu01: A C library for empirical testing of random number generators. ACM Transactions on Mathematical Software (TOMS) 33(4):22

L'Ecuyer P, Simard R (2014) On the lattice structure of a special class of multiple recursive random number generators. INFORMS Journal on Computing 26(3):449–460

L'Ecuyer P, Blouin F, Couture R (1993) A search for good multiple recursive random number generators. ACM Transactions on Modeling and Computer Simulation (TOMACS) 3(2):87–98

L'Ecuyer P, Simard R, Chen EJ, Kelton WD (2002) An object-oriented random-number package with many long streams and substreams. Operations Research 50(6):1073–1075

L'Ecuyer P, Nadeau-Chamard O, Chen YF, Lebar J (2021) Multiple streams with recurrence-based, counter-based, and splittable random number generators. In: 2021 Winter Simulation Conference (WSC), IEEE, pp 1–16

Lehmer DH (1951) Mathematical methods in large-scale computing units. In: Proc. 2nd Symp. on Large-Scale Digital Calculating Machinery, Harvard Univ. Press Cambridge, MA, pp 141–146

Lenstra AK, Lenstra HW, Lovász L (1982) Factoring polynomials with rational coefficients. Mathematische Annalen 261(4):515–534

Lewis TG, Payne W (1973) Generalized feedback shift register pseudorandom number algorithms. Journal of the ACM 20(3):456–468

Li CY, Chen YH, Chang TY, Deng LY, To K (2012) Period extension and randomness enhancement using high-throughput reseeding-mixing prng. IEEE Transactions on Very Large Scale Integration Systems 20(2):385–389

Li H (2005) A system of efficient and portable multiple recursive generators of large order. PhD thesis, University of Memphis, Memphis, TN., U.S.A.

Lidl R, Niederreiter H (1994) Introduction to finite fields and their applications. Cambridge University Press

Lüscher M (1994) A portable high-quality random number generator for lattice field theory simulations. Computer physics communications 79(1):100–110

L'Ecuyer P, Simard R (2013) TestU01: A Software Library in ANSI C for Empirical Testing of
Random Number Generators. University of Montreal

L'Ecuyer P, Munger D, Oreshkin B, Simard R (2017) Random numbers for parallel computers:
Requirements and methods, with emphasis on gpus. Mathematics and Computers in Simulation
135:3–17

MacLaren MD, Marsaglia G (1965) Uniform random number generators. Journal of the ACM 12:83–
89

Maitra S (2015) Chosen IV cryptanalysis on reduced round ChaCha and Salsa. IACR Cryptology
ePrint Archive URL http://eprint.iacr.org/2015/698

Maitra S, Paul G (2008) Analysis of RC4 and proposal of additional layers for better security margin.
INDOCRYPT 2008, Lecture Notes in Computer Science 5365:27–39

Mantin I, Shamir A (2001) A practical attack on broadcast rc4. In: Fast Software Encryption 2001,
Lecture Notes in Computer Science, Springer-Verlag

Marsagila G (1997) A random number generator for c. URL sci.stat.math

Marsaglia G (1968) Random numbers fall mainly in the planes. Proceedings of the National Academy
of Sciences of the United States of America 61(1):25

Marsaglia G (1972) The structure of linear congruential sequences. In: Zaremba SK (ed) Applications
of Number Theory to Numerical Analysis, Academic Press, pp 249–286

Marsaglia G (1985) A current view of random number generators. In: Computer Science and Statistics,
Sixteenth Symposium on the Interface. Elsevier Science Publishers, North-Holland, Amsterdam,
pp 3–10

Marsaglia G (1996) The Marsaglia random number CDROM including the DIEHARD battery of
tests of randomness. http://stat.fsu.edu/pub/diehard

Marsaglia G, Zaman A (1991) A new class of random number generators. The Annals of Applied
Probability 1:462–480

Marsland E (2011) Machine Learning. CRC Press

Mascagni M (1998) Parallel linear congruential generators with prime moduli. Parallel Computing
24(5):923–936

Mascagni M, Srinivasan A (2000) Algorithm 806: Sprng: A scalable library for pseudorandom number
generation. ACM Transactions on Mathematical Software (TOMS) 26(3):436–461

Massey FJ (1951) The kolmogorov-smirnov test for goodness of fit. Journal of the American Statistical
Association 46(253):68–78, URL http://www.jstor.org/stable/2280095

Matsui M (1993) Linear cryptanalysis method for des cipher. In: Workshop on the Theory and
Application of of Cryptographic Techniques, Springer, pp 386–397

Matsumoto M, Kurita Y (1992) Twisted gfsr generators. ACM Trans on Modeling and Computer
Simulation 2:179–194

Matsumoto M, Nishimura T (1998) Mersenne twister: a 623-dimensionally equidistributed uniform
pseudo-random number generator. ACM Transactions on Modeling and Computer Simulation
(TOMACS) 8(1):3–30

Matsumoto M, Saito M, Nishimura T, Hagita M (2007) A fast stream cipher with huge state space and
quasigroup filter for software, selected areas in cryptography. Lecture Notes in Computer Science
(LNCS) 4876:246–263

Matsumoto M, Saito M, Nishimura T, Hagita M (2008) Cryptmt3 stream cipher, new stream cipher
designs. Lecture Notes in Computer Science (LNCS) 4986:7–19

McCullough B (2006) A review of TESTU01. Journal of Applied Econometrics 21(5):677—682

McLeod AI (1985) Remark AS R58: A remark on algorithm as 183. an efficient and portable pseudo-random number generator. Journal of the Royal Statistical Society Series C (Applied Statistics) 34(2):198–200

Meier W, Staffelbach O (1989) Fast correlation attack on certain stream ciphers. Journal of Cryptography 1(3):159–176

Menezes AJ, Van Oorschot PC, Vanstone SA (1996) Handbook of applied cryptography. CRC press

Mironov I (2002) (not so) random shuffles of rc4. In: Yung M (ed) Advances in Cryptology - CRYPTO 2002. Lecture Notes in Computer Science, Springer, vol 2442

Mouha N, Preneel B (2013) Towards finding optimal differential characteristics for ARX: Application to Salsa20. In: IACR Cryptology ePrint Archive Report 2013/328

Mukherjee P (2013) An overview of estream ciphers. Centre of Excellence in Cryptology, Indian Statistical Institute, Kolkata, India

Nawaz Y, Gupta KC, Gong G (2005) A 32-bit rc4-like keystream generator. Cryptology ePrint Archive

Niederreiter H (1986) A pseudorandom vector generator based on finite field arithmetic. Math Japonica 31(5):759–774

Niederreiter H (1990) Statistical independence properties of pseudorandom vectors produced by matrix generators. Journal of computational and applied mathematics 31(1):139–151

Nir Y, Langley A (2018) Chacha20 and poly1305 for ietf protocols

Panneton F, L'ecuyer P, Matsumoto M (2006) Improved long-period generators based on linear recurrences modulo 2. ACM Transactions on Mathematical Software (TOMS) 32(1):1–16

Park SK, Miller KW (1988) Random number generators: good ones are hard to find. Communications of the ACM 31(10):1192–1201

Paul G, Maitra S (2011) RC4 stream cipher and its variants. CRC Press

Pearson K (1900) X. on the criterion that a given system of deviations from the probable in the case of a correlated system of variables is such that it can be reasonably supposed to have arisen from random sampling. The London, Edinburgh, and Dublin Philosophical Magazine and Journal of Science 50(302):157–175

Rivest RL, Shamir A, Adleman L (1978) A method for obtaining digital signatures and public-key cryptosystems. Communications of the ACM 21(2):120–126

Robshaw M, Billet O (2008) New Stream Cipher Designs: The eSTREAM Finalists, vol LNCS 4986. Springer-Verlag

Roos A (1995) Class of weak keys in the rc4 stream cipher

Sezgin F (1996) Some improvements for a random number generator with single-precision floating-point arithmetic. Computers & Geosciences 22(4):453–455

Sezgin F (2004) A method of systematic search for optimal multipliers in congruential random number generators. BIT Numerical Mathematics 44(1):135–149

Sezgin F (2006) Distribution of lattice points. Computing 78(2):173–193

Shannon CE (1948) A mathematical theory of communication. The Bell system technical journal 27(3):379–423

Shannon CE (1949) Communication theory of secrecy systems. The Bell system technical journal 28(4):656–715

Shchur LN, Butera P (1998) The ranlux generator: resonances in a random walk test. International Journal of Modern Physics C 9(04):607–624

Shi Z, Zhang B, Feng D, Wu W (2013) Improved key recovery attacks on reduced-round salsa20 and chacha. In: Information Security and Cryptology–ICISC 2012: 15th International Conference, Seoul, Korea, November 28-30, 2012, Revised Selected Papers 15, Springer, pp 337–351

Stallings W (2010) Cryptography and Network Security: Principles and Practice, 5th edn. Prentice Hall

Stallings W (2016) Cryptography and Network Security: Principles and Practice, 7th edn. Pearson

Stinson D (2006) Cryptography: theory and practice, 3rd edn. Chapman and Hall/CRC Press

Vanhoef M, Piessens F (2015) All your biases belong to us: Breaking RC4 in WPA-TKIP and TLS. In: 24th USENIX Security Symposium (USENIX Security 15), Washington D.C.

Von Schelling H (1954) Coupon collecting for unequal probabilities. The American Mathematical Monthly 61(5):306–311

Wagner D (1995) My rc4 weak keys. scicrypt, message-id 447o1l$cbj@cnnprincetonEDU

Watson EJ (1962) Primitive polynomials (mod 2). Math Comp 16:368–369

Wichmann B, Hill D (1982) An efficient and portable pseudo-random number generator. Journal of the Royal Statistical Society Series C (Applied Statistics) 31(2):188–190

Wu H (2004) A new stream cipher HC-256. In: Roy B, Meier W (eds) Proceedings of FSE 2004, Lecture Notes in Computer Science, vol 3017, Springer, pp 226–244

Wu H (2005) Cryptanalysis of a 32-bit RC4-like stream cipher. Cryptology ePrint Archive, Paper 2005/219, URL https://eprint.iacr.org/2005/219, https://eprint.iacr.org/2005/219

Wu PC (1997) Multiplicative, congruential random-number generators with multiplier $\pm 2^{k_1} \pm 2^{k_2}$ and modulus $2^p - 1$. ACM Transactions on Mathematical Software (TOMS) 23(2):255–265

Yao A (1982) Theory and applications of trapdoor functions. In: Proceedings of the 23rd Annual Symposium on the Foundations of Computer Science, IEEE Press, pp 80–91

Zeisel H (1986) Remark AS R61:a remark on algorithm as 183. an efficient and portable pseudo-random number generator. Journal of the Royal Statistical Society Series C (Applied Statistics) 35(1):89–89

Zierler N (1959) Linear recurring sequences. Journal of the Society for Industrial and Applied Mathematics 7(1):31–48